Making Time for Digital Lives

Making Time for Digital Lives

Beyond Chronotopia

Edited by Anne Kaun,
Christian Pentzold, and
Christine Lohmeier

ROWMAN & LITTLEFIELD
Lanham • Boulder • New York • London

Published by Rowman & Littlefield
An imprint of The Rowman & Littlefield Publishing Group, Inc.
4501 Forbes Boulevard, Suite 200, Lanham, Maryland 20706
www.rowman.com

6 Tinworth Street, London SE11 5AL, United Kingdom

British Library Cataloguing in Publication Information Available

Library of Congress Cataloging-in-Publication Data

Names: Kaun, Anne, editor. | Pentzold, Christian, 1981– editor. | Lohmeier, Christine, 1978– editor.
Title: Making time for digital lives : beyond chronotopia / edited by Anne Kaun, Christian Pentzold and Christine Lohmeier.
Description: Lanham : Rowman & Littlefield, [2020] | Includes bibliographical references and index. | Summary: "Explores the theory that digital technologies have multiplied and amplified inequalities rather than leveling the playing field as was promised and foretold"— Provided by publisher.
Identifiers: LCCN 2020024789 (print) | LCCN 2020024790 (ebook) | ISBN 9781786612977 (cloth) | ISBN 9781786612984 (epub)
ISBN 9781538149850 (pbk)
Subjects: LCSH: Digital divide. | Information technology—Social aspects. | Social change. | Digital communications.
Classification: LCC HM851 .M336 2020 (print) | LCC HM851 (ebook) | DDC 303.48/33—dc23
LC record available at https://lccn.loc.gov/2020024789
LC ebook record available at https://lccn.loc.gov/2020024790

Contents

List of Figures

List of Tables

Introduction

Making Time for Digital Lives: Sketching the Field and History of Resisting Dominant Temporal Regimes

Anne Kaun, Christine Lohmeier, and Christian Pentzold

Today's digital media and networked technologies have in general been associated with the temporal regime of immediacy, accessibility, and speed. Yet while they indeed seem to drive our sped-up lives, these notions of an increasing acceleration are inadequate to capture the complexity of lived time (Hartmann, 2019; Sharma, 2014). This volume gathers contributions that interrogate tactics to counter and resist that dominant form of temporality. They ask how digital temporalities are experienced and constructed, negotiated and transformed in modern timescapes. The volume aims to contribute to the growing realm of literature that engages with practices of disconnection, nonuse, and resistance to technologies and services, which seem to further a regime of speed. This regime can be understood both as discursive construction or moral imperative (Sutherland, 2019) *and* as temporal affordances that are emphasized, favored, and encouraged by digital media.

Disconnection practices from digital media that have also been called practices of technological nonuse, media resistance, or media disruption, have received an increased interest from researchers in recent years. Disconnection includes both "negative actions and attitudes towards media [and] describe[s] a refusal to accept the way media operate and evolve" (Syvertsen, 2017, 9). It thus moves into a different direction than earlier discussions of involuntary nonuse, focusing on cases when individuals are excluded from access based on social, economic, or infrastructural reasons. That way, nonuse or denetworking are not primarily a matter of inhibition or prevention but they have a deliberative stance. Rather than being a Luddite niche phenomenon or a sign of humanitarian and technological backwardness, abstention is a self-imposed choice. It becomes an evidence of self-care or self-determination, not a defect. This also implicates that the time and attention wrested from media should be put to some other creative, fulfilling, and meaningful purpose. Of course, the ability to make autonomous decisions to not use a communicative device or service is a privilege and cannot be enjoyed by all or at all times. Escaping an omnipresent mediated life is, after all, a fiction (Zurstiege, 2019). While not every moment of unplugging or media sabbath rests on rational decision-making, ideology, or deliberation, they require a position that allows to abandon a media offering or technology temporarily or for good. Hence, people rarely renounce media in total. In order to maintain social ties or be accessible for others, they might stop using some services but retain others. That way, media nonuse actually becomes a multifaceted and widespread field of differential media use.

Voluntary forms of disconnecting from digital media are currently often practiced in response to worries about addiction and overuse of digital technology that will have a negative impact on health and social life. Popular forms of disconnection in this regard include, for example, digital detox programs and camps that advocate digital health through digital fasting (Fish, 2017; Sutton, 2017). Other practices encompass nonuse and disconnection from specific content such as news (Strömbäck, Djerf-Pierre, and Shehata, 2012), suspending use of certain platforms (Brubaker, Ananny, and Crawford, 2016), or returning to "old" media emphasizing media nostalgia (Skågeby, 2011; Thorén and Kitzmann, 2015). Scholars have also highlighted the ambivalent role of disconnection practices for sustaining digital networks and how users in turn are drawn into the use of digital media while underscoring discourses of disconnection (Hesselberth, 2017; Light, 2014). Within the realm of disconnection studies, a number of researchers stress that reasons for and practices of media resistance are manifold and ingrained with habits and routines (Woodstock, 2014) as well as socially positioning (Wyatt, Graham, and Terranova, 2002). Often these studies of digital disconnection engage with individualized approaches of navigating digital culture today; there are, however, increasing attempts to conceptualize disconnection as collective, political practice (Kaun and Treré, 2018; Portwood-Stacer, 2013) and as a form of political asceticism against the dominance of network culture as Tero Karppi (2011) and Ulises Mejias (2013) suggest.

In the context of this growing area, the volume aims to answer what practices of digital disconnection tell us about the state of digital culture. What forms of agency are they implying and what alternative media cultures do they help us envision? These and related questions are addressed by this collection of case studies that move between conceptual pieces, research on digital disconnection, and synchronizing practices emcompassing both use and nonuse from a microperspective, as well as investigations into technologies of disconnection. Based on the collected effort put forward in the individual chapters, we formulate an argument of critical hope for our digital lives that moves beyond chronotopian and chronodystopian visions—suggesting what Ernst Bloch (1986) called a concrete utopia instead.

BEYOND CHRONOTOPIAS AND CHRONODYSTOPIAS—CRITICAL HOPE FOR OUR DIGITAL LIVES

This edited volume hopes to move beyond chronotopian and chronodystopian visions so to arrive at a more detailed discussion of digital disconnection practices. Instead of accentuating the empowering possibilities of disconnection or underlining the limits of media refusal, we consider them as expressions of critical hope that enact versions of a concrete utopia. We suggest to view practices of digital disconnection as forms of critical hope, critical hope that is political and that is not grounded in a naturalized idea of hope that people assume or that is linked to a natural historical development, but something that needs to be actively achieved and that is constantly negotiated. Critical hope that is expressed both through practices and actions as well as by way of critique.

In fact, critical theory rarely engages with the question of hope although, as Nicholas Smith (2005) argues, progressive critical theorists have presumably at least one thing in common: the hope for a better world. But other concepts have been judged more important and fruitful for critical inquiries, such as critique, rationality, justice, or memory. Hope, in comparison, has received comparatively little attention. Reasons for this disinterest might be that hope seems intellectually unsatisfactory, that we are hopeful in situations of uncertainty when we are not fully convinced or lack confidence. Hope, according to Nicholas Smith, is often connected with naïve and superficial optimism. Hope is also often linked to passivity as one is simply waiting for a hoped-for future and not actively pursuing and achieving it. But of course, hope is central for the motivation to pursue critical engagement with social reality. It is the ambivalence of hope that makes it such an interesting concept to consider. Hence, Gramsci (1989) reminds us in his prison notebooks of the importance of optimism of the will and pessimism of the intellect.

Although hope seems to be lacking an engagement by critical theorists, media and communication studies in turn has brought forward hopeful accounts of engaging with media. For instance, Natalie Fenton (2008) posits that hope in media studies is often conceptualized as agency through resistance. Take, for example, the seminal argument of the active audience and the idea that audiences resist the hegemonic representation in the text through active practices of appropriation, meaning-making, and interpretation that sometimes go against the grain and preferred or intended readings. Fenton thus writes,

> subcultures form acts of resistance displaying their profound disagreement with particular socio-political conditions in various ways. Journalists resist owner and editorial preferences through the sharing of collective professional values. Alternative media resist the frames, codes and practice of mainstream media through forms of organization, the means of production and modes of distribution. We look for resistance in every form of mediation and every act of consumption to satisfy ourselves and to kind of maintain our hope as media researchers in a better future. (2008)

In times where public discourse and academic research are focusing on the perils of digital technology such as the problems with delegating decision-making to algorithms, it remains important to also consider the agency of producers and users of digital media technology for the shaping of its outlook. This aspect is crucial for developing the kind of critical hope that we would like to stress throughout. This critical hope is a struggle that is ongoing and requires an active involvement and future visions based on a form of critical hope that does not assume a future that is already decided, but that is always still in the making. All chapters are examples of this kind of critical hope for media practices in digital culture that involve both connection and disconnection to enable media users and audiences rather than overemphasizing machine agency. The contributions zoom in on temporal aspects of these small acts of resistance to complicate our understandings of time in digital culture.

STRUCTURE OF THE BOOK

The book starts from the idea that power relations play out in time. Hence, people find themselves entangled in multiple temporalities where acceleration is not the universal tempo. While some enjoy the mixed blessings of a culture of speed, others have to carry its drawbacks, some of which are temporal in character again. Digital technologies and the social practices and wider economies they stipulate have, it seems, engrossed these discrepant politics of time. Hence, the publication charts different temporal trajectories. Going beyond time-based dichotomies of the fast versus the slow, the time rich versus the time poor, or a chronotopia versus a chronodystopia, its chapters show how people in diverse fields and situations cope with differential pace, manage the need for temporal coordination, and maintain or challenge their uneven and inequitable entimed relations. Of course, not all these endeavors are successful or can be sustained over a longer period. Some ventures to carve out temporal niches and to go against the tide might even seem outright quixotic. What they nevertheless indicate is the felt need to recuperate a kind of temporal autonomy and to assert the capability of spending time willfully.

These struggles for establishing differential times expand within the speed-up movement: they are about waiting and buffering, about slowing down, about the ironies of disconnecting through apps, or about the experiments with nonuse and media refusal. The volume thus invites us to appreciate the agency and ambition of people to deal with time and uphold temporalities that discontinue the hegemonic momentum and its demands of efficiency, timeliness, and attendance. Yet at the same time the multiplication of alternative chronographies also reveals the uneven prospects of working or actually not working on time. Far from being a smooth process, the empirically grounded studies in this book show how synchronization becomes political and is deeply marked by social inequalities playing out in time.

The chapters assembled in the book engage critically with alternative times that proceed besides acceleration. They thus hope to resist the moral, commercial, and cultural logics of the speed-up. With this, they assess a more versatile register of rhythms and dynamics that are both in and out of sync with the general real-time. The attempts to desynchronize and resynchronize digital lives presented in the contributions enfold the practical task of maintaining alternative temporalities. They often necessitate the adaptation of technological settings and they challenge the economic imperatives of a fast-paced modernity.

MAKING TIME FOR . . . DISCONNECTION

Disconnection has become a vivid topic for empirical and theoretical engagement. Despite the increasing number of publications, there is still a need to critically conceptualize what we mean with disconnection and disconnection from what (Hesselberth, 2017). The contributions by Tim Markham as well as by Ingrid Forsler and Carina Guyard offer ways in which we can think of disconnection as a practice and a discourse. They suggest an analysis of negotiating disconnection in relation to connection as a dialectic process that is continuously taking shape through practices and discourses. Magdalena Kania-Lundholm engages with disconnection from the perspective of elderly seldom-users of digital technology. She provides answers to the questions of what it is that they lose and what is it they gain from not using certain digital services and how they relate to the discourse of the digital imperative. Christian Schwarzenegger and Manuel Menke, combining several connate studies, consider how media users are creating pockets of temporal autonomy by way of disconnection and nonuse while remaining connected through digital media. All four contributions conceptually and empirically widen and enhance the notion of disconnection. They remind us of the ambivalent character of nonuse in lives never fully disconnected.

MAKING TIME FOR . . . SYNCHRONIZATION

Claiming temporal autonomy often takes the shape of attempts to synchronize different tasks and practices. Martin Hand, for example, illustrates how his respondents orchestrate their lives and the lives of their family members, colleagues, and employees with the help of digital technologies. Here, disconnection and nonuse are integrated in busy schedules that are largely depending on digital technologies. The smartphone becomes especially prevalent for these synchronization practices. Roxana Morosanu Firth, Sean Rintel, and Abigail Sellen zoom in on how smartphones allow for temporal mobility and multitemporality. The smartphone thus becomes a tool that captures and produces diverse temporalities simultaneously and on the move. In their chapter, Hannah Ditchfield and Peter Lunt examine communicative practices of online interactions that are reconfiguring synchronicity and sequentiality. They show how multicommunication, that is, communicating with different people at the same time and on different levels, requires particular ways of synchronizing different communicative engagements. This, too, includes waiting and managing impatience as elements of conversational disconnection.

MAKING TIME FOR . . . COMMODIFICATION

Disconnection takes shape not only in practices and discourses of nonuse but is also increasingly commodified. A plethora of applications and widgets offer us ways to disconnect and regain focus and creativity. While Alex Beattie as well as Mikolaj Dymek look at specific applications, Siempo and OmniFocus, respectively, Carla Ganito and Cátia Ferreira provide a broad overview over different disconnection applications from a comparative perspective. All these products share the basic assumption that connectivity is the default but that users need to carve out spaces of disconnection, preferably with a digital application.

Through the three areas of disconnection as practices and discourses of nonuse, as practices of synchronization and orchestration, and as commodification, the volume demonstrates the complex and creative ways in which users make time for their digital lives that move beyond simplistic dystopian and utopian assumptions of digital connectivity.

REFERENCES

Bloch, E. (1986). *The principle of hope*. Cambridge, MA: MIT Press.

Brubaker, J. R., Ananny, M., and Crawford, K. (2016). Departing glances: A socio-technical account of 'leaving' Grindr. *New Media & Society, 18*(3), 373–90.

Fenton, N. (2008). Mediating hope: New media, politics and resistance. *International Journal of Cultural Studies, 11*(2), 230–48.

Fish, A. (2017). Technology retreats and the politics of social media. *TripleC, 15*(1), 355–69.

Gramsci, A. (1989). *Selection from the Prison Notebooks*. London: Lawrence & Wishart.

Hartmann, M. (2019). The normative framework of (mobile) time: Chrononormativity, power-chronography, and mobilities. In M. Hartmann, E. Prommer, K. Deckner, and S. O. Görland (Eds.), *Mediated time: Perspectives of time in a digital age* (pp. 45–66). London: Routledge.

Hesselberth, P. (2017). Discourses on disconnectivity and the right to disconnect. *New Media and Society*, 1–17.

Karppi, T. (2011). Digital suicide and the politics of leaving Facebook. *Transformations* (20), 1–18.

Kaun, A., and Treré, E. (2018). Repression, resistance and lifestyle: Charting (dis) connection and activism in times of accelerated capitalism. *Social Movement Studies*, 1–19. doi:10.1080/14742837.2018.1555752

Light, B. (2014). *Disconnecting with social networking sites*. Basingstoke: Palgrave Macmillan.

Mejias, U. A. (2013). *Off the network: Disrupting the digital world*. Minneapolis: University of Minnesota Press.

Portwood-Stacer, L. (2013). Media refusal and conspicuous non-consumption: The performative and political dimensions of Facebook abstention. *New Media & Society, 15*(7), 1041–57. doi:10.1177/1461444812465139

Sharma, S. (2014). *In the Meantime: Temporality and cultural politics*. Durham, London: Duke University Press.

Skågeby, J. (2011). Slow and fast music media: Comparing values of cassettes and playlists. *Transformations* (20), 1–17.

Smith, N. (2005). Hope and critical theory. *Critical Horizons*, *6*(1), 45–61.

Strömbäck, J., Djerf-Pierre, M., and Shehata, A. (2012). The dynamics of political interest and news media consumption: A longitudinal perspective. *International Journal of Public Opinion Research 25* (4), 414–35.

Sutherland, T. (2019). The categorical imperative of speed: Acceleration as moral duty. In M. Hartmann, E. Prommer, K. Deckner, and S. O. Görland (Eds.), *Mediated time: Perspectives on time in a digital age* (pp. 25–44). Cham: Palgrave Macmillan.

Sutton, T. (2017). Disconnect to reconnect: The food/technology metaphor in digital detoxing. *First Monday, 22*(6). doi:http://dx.doi.org/10.5210/fm.v22i6.7561

Syvertsen, T. (2017). *Media resistance: Protest, dislike, abstention*. Cham: Palgrave Macmillan.

Thorén, C., and Kitzmann, A. (2015). Replicants, imposters and the real deal: Issues of non-use and technology resistance in vintage and software instruments. *First Monday, 20* (11). doi:http://dx.doi.org/10.5210/fm.v20i11.6302

Woodstock, L. (2014). Media resistance: Opportunities for media practice and new media research. *International Journal of Communication, 8*, 1983–2001.

Wyatt, S., Graham, T., and Terranova, T. (2002). They came, they surfed, and they went back to the beach: Conceptualizing use and non-use of the Internet. In S. Woolgar (Ed.), *Virtual society? Technology, cyberbole, reality* (pp. 23–40). Oxford: Oxford University Press.

Zurstiege, G. (2019). *Taktiken der Entnetzung. [Tactics of De-Networking].* Berlin: Suhrkamp.

Part I

MAKING TIME FOR . . . DISCONNECTION

Chapter One

Subjective Recognition in a Distracted World

The Affordances of Affective Habits and Temporal Discontinuities

Tim Markham

In an era characterized by an economy of attention (Franck, 2019), the temporal resources of individuals are regarded as a finite resource. Distraction is the norm with the digitally saturated environments we inhabit presenting stimuli far in excess of what anyone could ever hope to process and respond to. This has practical consequences for those in the business of representing the distant suffering of others to media consumers, as well as to advertisers faced with the impossibility of targeting an attentive, captive audience. Underlying these there are more resonant questions about how we pay attention to a world that always exceeds our capacities, questions that long predate the digital age. This chapter places temporality front and center in order to investigate how intersubjective recognition can and should function in everyday digital life. In particular, it argues against the instinctive response that the key to recognition is either temporal extensiveness or intensiveness—that is, a long, unremitting glare at another (Buber, 1942), or an intense focus that ruptures ordinary conscious experience (Sontag, 2013). It argues instead that intersubjective recognition that is both ethical and sustainable is produced through and not in spite of the ordinary, affective distractions of quotidian existence and the habituated, instinctive practices we enact instinctively as we make our way around digital environments at work or leisure. The chapter is based on insights gleaned from semi-structured interviews with twenty heavily digitally embedded journalists in Egypt and Lebanon, though it is argued that the findings have broader resonance. Originally conducted in order to assess how such individuals experience moments of great historic change, it emerged that the way they inhabited digital spaces speaks to more universal themes—namely, that political principles and personal motivations are not internal qualities wielded upon the world, but the outgrowth of patchy, temporally provisional digital habits that we find ourselves doing amid the mundane rhythms of everyday life.

THEORETICAL FRAMEWORK

The theoretical framework underpinning this work combines Emmanuel Levinas's notion of pure duration (1996 [1976]) with Martin Heidegger's exegesis on idle talk, curiosity and ambivalence in *Being and Time* (1963 [1927]). It is fair to say that Levinas is often overlooked in media and digital scholarship, though his innovations have been much revived by the likes of Amit Pinchevski (2005) and Paul Frosh (2011; 2018). His principal provocation was to invert Aristotle's assertion that ontology is the core of the philosophical enterprise, asserting instead that ethics is the "first philosophy."[1] What this boils down to is that ethical relations between humans is not predicated on their knowing each other, but on the initial inkling of the brute fact of their coexistence and mutual complicity and responsibility *before* these things are made conscious objects. As in all phenomenology, existing is prior to existence: we do not start as discrete subjects that emerge to engage with the world; instead we are always, already active in it, doing things and subjectivating with whatever resources are at hand. To make others an object of knowledge or emotion—compassion or pity, say—is an act of violence; all objectification proceeds through the effacement of difference, of making the other identical to myself.[2] Pure duration does not denote the gradual accumulation of awareness of the alterity of those we discover ourselves amongst, but rather the preconscious zone in which that alterity eludes us and asserts itself. This then opens up the ethical affordances of intersubjective relations that are merely felt rather than cogitated, peripheral rather than concentrated, distracted rather than focussed.

Levinas thus brackets out the metaphysical aspects of a quality such as solidarity: there is no "deep" revelation to be excavated concerning what is means to be in the world with others, instead insisting that the meaning of relationality consists only in the way we navigate worlds into which we are thrown through improvisation, learned shortcuts, and by feel. The kind of non-intentive consciousness in which pure duration consists can sound like a form of willed naivety, and in the more colourful passages of *Otherwise than Being* (2013 [1974]) it has spiritual overtones, but its efficacy as a concept is much more practical. Solidarity is not an essential quality or something that an individual must definitively decide to live by, but a mode of relationality entangled in the practices of existing we find ourselves engaging in before "we" exist. The extent to which someone is solidaristic or compassionate towards multitudinous others is not innate, and nor does it have to be felt passionately. It does, though, matter to what extent it feels intuitive—and that it is because it is the product of habituation over time. In Katherine Withy's (2015) reading of Division Two of *Being and Time* she uses the term disclosive posture to encapsulate the idea that selfhood is radically contingent, insistently present and provisional, yet sustainable *as* selfhood as ways of making the world and the self familiar solidify into orientations or bearings.[3] Endless improvisation produces contextually specific response templates, and that inevitably means that individuals come to encounter others generically and heuristically. The upshot is that the normal mode of being in the world is one in which we fail to recognize others for who they are, and we fail to communicate with them, most of the time. The twist is that whereas for Sartre (1992) there is tragedy in such ordinary failure, for Levinas it is a promising start: "The failure of communication is the failure of knowledge"—and this is a good thing. As Pinchevski (2012, 356) puts it, "[i]t is precisely in moments of uncertainty and instances of misunderstanding, lack, or even refusal that I find myself facing the other."

The idea that failing to recognize others for what they are is not just forgivably human but the basis for a uniquely ethical relationality seems on the face of it counterintuitive, but it chimes with the phenomenological claim seen in Heidegger that the primary function of perception and communication is not to impart knowledge or convey mental states from one mind to another, but instead "intersubjective world sharing"—an affirmation of being with (and potentially for) others. Frosh (2011, 393) has put this to good effect in his work on news audiences, arguing that their unremarkable indifference to the strangers encountered there, to their "constant and cumulative presence within the home," should be seen as a historical accomplishment. Finding the right ideas and words accurately to explain the experiences depicted on domestic screens is an example of what Levinas terms the Said—language used intentionally to represent and designate; what is of more interest to Levinas is the Saying, which is about our preverbal exposure and openness towards the other.[4] This has real implications for digital research, because it suggests that rather than asking "digital existers"[5] what they think about this or that experience or this or that digitally rendered other, the point becomes to take seriously the way that such encounters are flattened out until they barely register at all, serving merely as something that moves one's attention from one thing to the next. It is the paths and rhythms of such movements in the temporal experience of everyday digital life that afford the possibility of the kind of ethical interrelationality that Levinas proposes.

FIELDWORK IN EGYPT AND LEBANON

It would be unfair to suggest that the journalists in Cairo and Beirut lacked passion or knowledge about the others they encountered in their work, with interviews expressing expertise and finely tuned analyses of everything from the war in Syria and the Arab uprisings of the early 2010s to endemic political corruption, poverty, and the influence of Hezbollah and Hamas. There is also no denying the perils faced by media professionals in the Middle East, from censorship and imprisonment to sexual harassment and politically motivated violence. It would be a mistake, however, to infer that they persevere in their work because of the kinds of individuals they are; if we take the phenomenological model to its logical conclusion then they are the kinds of individuals they are because of the persevering they find themselves doing. All respondents could readily invoke a personal sense of professional mission and alternately dystopian and utopian projections of what the future might bring. But the headline conclusion that this research produced was that their motivation, their sense of responsibility to those whose miseries and injustices they witnessed and documented, was secured at a different temporal order. It was, in short, about the things that made things meaningful at the everyday level, things that on the face of it seem banal, frivolous, even inappropriate to the task at hand.

Specifically, these took the form of the digital habits that did not so much get in the way of the seriousness of their work as provide the impetus that moved them from one moment to the next across their busy days. Getting at this level of experience was a methodological challenge, with some scratching of heads resulting when they were asked how soon after waking up they checked in on Facebook and Twitter and how they dealt with their email inboxes on arriving at the office. But it soon became clear that the endless posting, liking, and responding on social media platforms, checking in on traffic updates, texting potential sources about meeting and friends about where to go for dinner, responding to insistent emails from bosses about reach and impact all combined into a hum that was less a matter of persistent intensity and more about improvised, provisional activity. The distractions and interruptions, unexpected developments and changes of plan provided temporal momentum rather than hindering it. Journalists are no different from doctors or councillors in that those to whom they owe a duty of care necessarily become a matter of routinization and logistics, tasks to be dealt with. But this way of living, which applies equally to ordinary digital citizens, is also what makes it just-liveable—that is no say, not *endurable* but ordinarily meaningful. Otherwise it was the affective experience of one thing after another—the way the interview's tone changed when describing the frustrations of receiving an email from management or a bit of gossip from a colleague, and especially monitoring the number of responses to an article they wrote and the way they responded in turn—that moved things along, disclosing the stakes that kept them in the game.

The implication then is that the undoubted motivation and principle possessed by respondents were not instantiated in spite of the practical and affective distractions of everyday life, but precisely through them. The experience of busyness in everyday digital life matters just as much as an ultimate goal towards which someone might be aiming; the constant felt sense of activity suggests a temporality that is much more about feeling one's way from one encounter to the next *without* attending to longer term temporal horizons. Pivotally, this is not about a gap between the personal and political, a disconnect between the kinds of bearing or disclosive postures demanded by each. The ambivalence and provisionality of personal experience is what sustains political engagement and efficacy, since it affords a non-objectifying, non-intentive relation towards the stuff of the latter that is genuinely open, exposed, and culpable.[6] Likewise, the sense of camaraderie on display in a newsroom is by this logic not the product of like-minded, similarly principled souls coming together; rather, the affective pleasures (and displeasures) of ordinary workplace socializing, notably by way of digital platforms and devices, are what gird that sense of collective endeavour. The palpable buzz felt by many respondents on seeing their words and videos published on their organization's website, not to mention when they dive into the below-the-line melee, might seem like an unbecoming, egotistical distraction from the seriousness their role demands. But seriousness is neither innate nor acquired; it has to be sustained temporally by movement through everyday worlds, and it is more than plausible that the hot flush of digital status recognition can provide the necessary fuel.

IN DEFENSE OF AFFECTIVE DISTRACTION

Frosh has developed a compelling critique of the attentive fallacy: "the assumption that the significance of representations is generated through an intense, focussed interchange between an attentive address and a formally distinct, unified text" (2018, 13). The same line of argument can be applied to any kind of digital encounter: Why should attention and focus be privileged over the ambient, peripheral, ephemeral, and fragmented? That we have rapidly developed ways of experiencing things digitally with such effortlessness, before seamlessly turning our gaze towards whatever comes next, ready and even eager to instinctively size it up and respond or not as the situation demands, and on and on, could be seen as ingenious rather than somehow lacking. Ben Highmore (2010) has similarly made the case for the defence of distraction as a kind of "promiscuous absorption," making the point that it is not just harmlessly human, but a restless agility that has real ethical affordances. The impatient darting about that is associated with cultures of digital media practice—the constant checking in, tapping, and swiping—can thus be seen as a uniquely appropriate means of populating the digital worlds we inhabit with the bodies and voices of others. The non-intense relationships subjects have with other subjects, not to mention objects from devices to infrastructures, provide the basis for a practicable ethics of being-in-the-world. As Frosh (2018, 45) puts it, it is indifference acting as a moral force in the taken-for-granted, pre-reflective experience of everyday life.

This points to a different approach to temporality than is usually presumed in debates around digital ethics. Carolyn Pedwell (2017), drawing on the geographical phenomenology of Tim Ingold (2000),[7] argues that what is required is an experience of time without movement: that is, we need to "inhabit affect," to pause, dwell, and reflect on how we feel amid digital environments, what is causing us to feel that way and what our obligations to others known and unknown are. Pedwell is emblematic of a school of thought that holds that an appreciation of the contingencies and ethical implications of the digital worlds we live in can only be achieved through rupturing the otherwise seamless everyday experience of it, pulling back to coolly assess what is really going on and what is at stake. In fact, Ingold is skeptical of deriving much of anything from a bird's-eye perspective: we live longly, not vertically; that is not a limitation but how the world is disclosed to us, and as such the basis of reality itself. And while it is possible to think about time without movement, movement without time is an impossibility. If we accept that ethics only makes sense temporally, concerning the differential futures that could result from what happens by design, accident, or default in the present, then all movement through digital spaces, however mundane, cannot help but be ethical in nature.

In joining the dots between distraction and absorption, Highmore is following in the footsteps of both Walter Benjamin and Siegfried Kracauer. The common thread between the three is a commitment to the specifically embodied kind of knowledge enabled by distracted motion from one object to another. Importantly, the distracted gaze does not require any special effort, imagination, or creativity; it is simply the humdrum work of making sense of the everyday world. Thus, while the journalists interviewed are likely better informed about the detail, the history, and the politics of the issues on which they report, there is nothing particularly skillful about the way they engage in digital practices; their tics and quirks, habits and affective responses are much like anyone else's in terms of the knowledge they afford. Expertise matters greatly: respondents' knowledge of human rights law and the convoluted, sclerotic factionalism that cuts across the ordinary lives of citizens in both Egypt and Lebanon shed light on the challenges they face in their work as well as their motivations. But by taking Levinas seriously it is possible to get at a different level of knowledge, that gained in the thick of it, in digital environments peopled by the others of their work—the poor, the corrupt, the stateless—who are encountered barely more than as points in time and space. Fleetness of foot in their digital wayfaring does not make these practitioners callous; it is precisely what incarnates their motivations and principles. As Kathleen Stewart (2007) has argued previously, it requires only the barest affective hit to propel the digital inhabitant forward, so whatever provides that propulsion can be thought of as ethically constitutive.

For Highmore the key to distraction is its unresolvedness, its openness matched by an orientation towards fixing one's gaze that acts as an inexhaustible resource of mobility. As well as a lack of resolution, however, we could add provisionality. The distinction between the two is that provisionality entails a position-taking in relation to objects encountered; each act of glancing, responding (or not), and moving on has attendant stakes—there are costs and payoffs, risks and rewards that matter, however minutely. Levinas is careful to remind the reader that the state of pure duration is essentially accusatory; there is nothing like wide-eyed wonder in making one's way about the world. Benjamin's account of distraction has a leisurely, edifying tone, but for Levinas the knowledge one acquires through this non-intentive navigation consists in culpability more than anything else. Provisionality then is not a matter of dilettantism in ethics or subjectivity, but it does mean that each of these should effectively be conceived as exploratory and experimental. If subjectivation is by definition unresolved, a process rather than destination of becoming, then it makes sense that provisional position-takings will be refined and discarded as we go. That which sustains constancy over time is neither an eminence from within nor a declaration of purpose[8] but the repertoires developed individually and collectively for responding on the fly. Inhabiting digital spaces through affective motion is an unrelenting disclosure of the world and of one's relation to it. Rather than casting such spaces as sideshows designed to keep us occupied so as to avoid thinking about what "really" matters, it is in the deftness and fluidity of our experience of them that manifests what is at stake.[9]

HEIDEGGER ON PASSING THE WORD ALONG

Momentum is a collective endeavour. Much of what interviewees did by way of engaging digitally across daily life involved regular, unlaborious contributions to keeping discussion going around whatever topics they were covering: noticing what else was being said, responding affectively as well as through further written interventions and the usual panoply of social media shorthand. None would claim that any particular enunciation "really" grasped the reality on the ground; instead, as Heidegger puts it, "we already are listening only to what is said-in-the-talk as such" (1963, 212). This is inauthentic talk to be sure, talk which happily hoovers up any and all discussion points and readily understands them, albeit usually approximately and superficially. It is at this point, though, that Heidegger inserts that claim about communication being more an expression of being-with than of imparting knowledge. Inauthenticity is really beside the point: "the fact that talking is going on is a matter of consequence" (1963, 212). Invariably the way we discuss things in everyday life, even in a professional capacity, will have long lost its primary "relationship of being" towards the thing itself. But this discoursing does not function in order to get at such a primordial apprehension of being; it "communicates rather by following the route of *gossiping* and *passing the word along*" (emphasis in original, 1963, 212). Heidegger calls this idle talk, but as with inauthenticity, he is at pains to emphasize that this is not meant to be dismissive. In an ontological sense, idle talk is as factual, as grounding of the real, as anything else—*because* it projects forward in time.

There is something almost playful in the way that Heidegger writes of everyday talk that it dispenses with the need to distinguish that drawn from primordial sources and mere gossip: it "does not need it because, of course, it understands everything" (1963, 212). Importantly, the ability to discuss this and that without making them one's own gives idle talk a groundlessness;[10] it has been "uprooted existentially, and this uprooting is constant" (1963, 214). Dasein in this state floats along unattached to objects in the world; "yet in so doing, it is always alongside the world, with Others, and towards itself" (1963, 214). He then goes on to define curiosity as the prioritization of seeing over sight, without being restricted to an orientation towards cognition. Curiosity has no intention of really understanding what it sees; it is only interested in seeing—to which one might add: What is seeing but sight made temporal? The condition of seeking out novelty and liveliness is at the heart about the possibilities of abandoning oneself to the world. It does not have to be about marvelling at what one encounters; instead is about knowing "just in order to have known" (1963, 216). By combining the desires not to tarry and to be distracted by new possibilities gives curiosity the character of "never dwelling anywhere" (1963, 216). And the point is that this does not take the subject ever further away from Dasein, for never dwelling anywhere is itself ontogenic, constitutive of the real. Curiosity is thus "everywhere and nowhere" (1963, 217), but reveals a new kind of being of everyday Dasein characterized by its constantly uprooting itself.

A few pages later Heidegger expands on how idle talk is temporally distinct and thus singularly disclosive of Dasein and its being in the world with others. When Dasein pauses to reflect, to make an effort to really understand what an object *is*, "its time is a different time"—namely, slower. Idle talk, on the other hand, "lives at a faster rate" (1963, 218). While Dasein cogitates, idle talk will have long since moved on to the next thing, the next discussion point. There is an ambiguity [*Zweideutigkeit*] at the core of this movement, in that Dasein is always unsure when engaged in discourse whether the thing being discussed is really the thing being discussed, but at the same time curiosity "gives idle talk the semblance of having everything decided in it" (1963, 219). There is something about this ambiguity that particularly resonates with the experience of lively social media exchanges, in that it feels like everything is up for grabs, that things can be made sense of and resolved, and yet there remains the nagging doubt that this is simply talking for the sake of talking. One of the Beirut reporters described as "compulsive" his constant arguments with interlocutors on Facebook, and freely admitted that these debates were often "ridiculous" while concluding that "You can't stay detached from it, you have to be in the matrix." As was seen earlier in terms of distraction, the ambiguity of curiosity is its simultaneous lack of resolution and promise of the same: *that* is what provides momentum.

It is tempting when Heidegger writes of "fallingness" to imagine a fall from grace, a corruption of pure Dasein by the grubby business of living in the world. His point, however, is that fallingness is always already our starting point. To find oneself fallen into the world means to find oneself in a state of absorption with being-with-others—and this is "guided by idle talk, curiosity and ambiguity" (1963, 220), not occluded by them. Fallingness implies movement, and movement implies temporality—there is no Dasein outside of time. Everyday digital chatter might be inauthentic in the sense that it is motivated by the desire to move things along more than by wanting to nail down the nature of the real, yet it nonetheless discloses to Dasein a mode of being towards the world as well as towards others and towards itself. These things may be understood to a degree, but not in a way that would stand up to much scrutiny. Curiosity discloses everything, and ambiguity hides nothing from Dasein, but the fact that all is grasped from a position of groundlessness means that there is never an "aha" moment. This, then, is what Frosh (2018) criticizes when he writes of the "rupture of rapture," seen in many accounts of digital ethics in which the assumed goal is some kind of revelation as to what is really going on. If we follow Heidegger's logic, the kind of knowledge we pick up on the run in a constant state of restless uprootedness, full of doubt and promise as to what is known and what can be known, is a more promising route for grasping being in the world.

In temporal terms, the hum of activity at work in general and through constant engagement in digital media environments can seem like a displacement strategy, a way of suspending taking decisive action towards a determinate conclusion. Heidegger gives a good sense of this suspicion here, the way we collectively convince each other that our discourses and routines are meaningful giving us the comforting excuse of remaining in a holding pattern. It is arguable that this is always a risk in any work environment, with routinization and bureaucratization always geared towards sustaining day-to-day meaningfulness and over time coming to serve *only* that purpose. The same might be said of our social and romantic relationships, after all, with existential reservations about their meaningfulness deferred by the lower intensity whirr of shared rituals. The research behind this chapter originated in that dilemma all professionals face: In the absence of undeniable climaxes of achievement, what beyond material needs provides the basic motivation to wake up, travel to work and do the same thing day in, day out? One of the main findings was that work routines cannot be dismissed as a smokescreen to avoid facing up to the existential meaninglessness of all human endeavor; more than a coping mechanism, they are what make abstractions like principle and motivation a reality in the world. That is, they disclose, and not occlude, the meaning of such ideas.

By zooming in specifically on the digital habits of respondents, the same logic can be seen to apply. It is easy to write off diving into a Twitter storm or WhatsApp flurry as what Heidegger calls tranquilization, discourse that is sufficiently familiar and affectively intense as to justify its own meaningfulness indefinitely—passing time in a way that feels just substantial enough without demanding actual intersubjective recognition and responsibility. But as with pure duration, "this tranquillity in inauthentic being does not seduce one into stagnation and inactivity" (1963, 222). For one of the Lebanese journalists, checking Facebook and Twitter every twenty minutes clearly served this function, saying that it was not really about gathering information but keeping her "awake and motivated." Indeed, the "hustle" to which we abandon ourselves is what prevents being-in-the-world from coming to rest; it aggravates the state of being-fallen towards—well, precisely what remains at this point unclear. There is certainly no guarantee that the aggravation of falling will lead to Dasein finally gaining enlightenment about the world and its place within it, and it remains the case that "versatile curiosity and restlessly 'knowing it all' masquerade as a universal understanding of Dasein" (1963, 222). The point is that there is no way of knowing what there really is to be understood, no sense in holding out that the true meaning of that in which we are busily engaged will reveal itself as itself. The world is disclosed forever incrementally; whatever propels us forward in time through that world reveals its meaningfulness as we go. The constant dipping in to chatter about the things we are interested in or simply distracted by cannot be judged by how close it gets us to some ultimate moment of realization and subjectivation, only by what where it leads us next and with what implications and affordances.

None of which is to suggest that every fit of moral pique or elaborate performance of virtue signalling on a digital platform is a good thing. For Heidegger there is always the possibility, probability even, that busying oneself with discourse will lead to alienation, described as Dasein "getting tangled in itself" (1963, 223). At the end of Division One he underlines the point that authenticity is not something that somehow exists separately from the lived everyday experience of being-in-the-world, and fallingness does not give us a kind of "night view" of Dasein. As he puts it, "Falling reveals an essential ontological structure of Dasein itself. Far from determining its nocturnal side, it constitutes all Dasein's days in their everydayness" (1963, 223). It follows that the ethical parameters of any digital space are constituted in motion across time; they emanate from all the stuff that happens within it and drives it forward, not in spite of it. More specifically for the respondents in this study, it means that their relationality towards those they see their work as being about and for—the otherwise voiceless and powerless members of society as well as their audiences—is instantiated in the sometimes frenetic, often distracted, usually ambivalent way that they go about their everyday routines, and not in those brief periods when they are actually interviewing sources or reflecting at leisure about their responsibility to their publics.

For Heidegger, the world is disclosed moodwise. This means that the temporal experience of digital environments with feelings of busyness, distraction, pleasure, indignation, urgency, boredom and ambivalence should not be thought of as a subjective layer that sits on top of and obscures the hard reality beneath. The world made real through such states is the base line of facticity; it is always the point of departure of phenomenological investigation rather than a diagnosis of a defect to be corrected, the inescapable affective noise of the mortal realm that somehow has to be filtered out. Further, these states do more than give us the occasional glimpse behind the curtain; there is no backstage to be revealed, only what is instantiated through being in and navigating the world. There is no ultimate revelation at some imagined end of being in the world with others, nor any claim to reclaim an originary solidarity long since compromised by the depravities and degradations of modern life.

In research into digital temporalities, this frees us from the commitment to imagining new kinds of representation or mediated experience that will break through the cyclical fog of our overstimulated everyday lives to clinch real subjective recognition of the other. In professional contexts that other is objectified through work routines to become temporally specific tasks experienced affectively, but this does not preclude substantive subjective recognition. By this logic, the other encountered digitally, often experienced fleetingly and similarly affectively, need not be reduced by its digital mediation to inert spectacle. Under particular circumstances it is possible that temporal, affective digital habits can underpin more politically substantial relationships between subjects in the longer term. The question then becomes how do we critically scrutinize both digital environments and individuals' engagement with them.

CONCLUSION

The current research concluded that *asking* participants how they went about their digital activities and why was of limited use. All were able spontaneously to express principles and goals, tactics and strategies; the point is that these invocations are not expressions of inner subjectivities but subjectivation in action, ways of disclosing and becoming that are eternally in the present tense. It is a matter of comparing and contrasting their everyday paths of wayfaring digitally, regardless of whether these are the result of conscious decisions or arbitrary habituation. What are the constraints and affordances of tracing this path repeatedly over time compared with that one? If not these routines of engaging digitally, what else? The differences are real enough, with those inhabiting affective loops of heat and fury, relishing fighting the good fight, not just having different attitudes to those quietly working through to-do lists of scheduled tweets and fact checking, but as a result existing differently in relation to the others of their work. In both cases those intersubjective others encountered on the fly in an ordinary day's work became a sideshow to their conscious intentiveness, but the point is that this withdrawal to the periphery of consciousness might just be the ideal ethical position-taking in relation to them.

There may be broader implications to tease out from the insights into this particular group of media professionals. Inhabitation of digital spaces in time is about making it familiar through habituation. Repetition is a necessity not a hindrance, though that which it constrains and affords matters in the very real sense of how it discloses the world and one's place in it. The way some respondents labored through daily cycles of inbox clearing and hashtag monitoring made clear that being in a rut can stifle one's motivation and efficacy—but the only alternative is to be in a different rut, to carve out a new path through new habits of digital locomotion. Forming new digital habits need not be the result of a revelatory spark, simply trying out new things and seeing where it goes. Curiosity is the key here, and it is clear that the ability to explore and experiment, free from the nagging question of whether it is authentic or mere distraction, is an important one—for all digital existers. Skills of improvisation are also important, and these depend on the application of acquired repertories that come more naturally to some than others. It is then just conceivable that practical digital facility and agility are of greater importance than digital literacy conventionally conceived. Finally, the participants highlighted the provisionality of digital existence in time: today's arguments would be forgotten tomorrow, the casual yet constant checking-in on social media rarely yields knowledge they would directly incorporate in their work, and yet it was all palpably worth doing, and doing so with energy and determination. All were invested in their digital engagement, their embodied commitment to their ways of going about it amounting to a practical knowledge of what is at stake. Provisionality shows that improvisation and commitment are not incompatible, the former revealing the contingency of the latter. The result for all digital inhabitants is that restlessness and an appetite for novelty is not necessarily a sign of inconstancy. What it means to exist ethically in relation to others in a digital world is not produced by standing back from it, but getting stuck in it.

NOTES

1. Sean Hand, ed. *The Levinas Reader* (Oxford: Wiley, 2001).
2. "One does not see that the success of knowledge would in fact destroy the nearness, the proximity, of the other." Emmanuel Levinas, *Otherwise Than Being or Beyond Essence* (New York, NY: Springer, 2013 [1974], 103–4).
3. See also Jan Slaby and Gerhard Thonhauser, "Heidegger and the Affective (Un)Grounding of Politics," in *Heidegger on Affect*, ed. Christos Hadjioannou (Basingstoke: Palgrave Macmillan, 2019).
4. See especially Levinas, *Otherwise Than Being,* 120.
5. Amanda Lagerkvist, "Existential Media: Toward a Theorization of Digital Thrownness," *New Media & Society* 19: 1 (2017): 96.
6. Levinas, Otherwise Than Being, 48.
7. See also Shaun Moores, "Digital Orientations: 'Ways of the Hand' and Practical Knowing in Media Uses and Other Manual Activities," *Mobile Media & Communication* 2: 2 (2014): 196–208.
8. This echoes Heidegger's notion of "standing in a situation."
9. This is adjacent to Pierre Bourdieau's *Sens pratique* (see Bourdieau, 1988: 25).
10. See also Slaby and Thonhauser, "Heidegger and the Affective (Un)Grounding of Politics."

REFERENCES

Buber, Martin. *I and Thou*. Edinburgh: T&T Clark, 1942.

Franck, Georg. "The Economy of Attention." *Journal of Sociology* 55, no. 1 (1 March 2019): 8–19.

Frosh, Paul. "Phatic Morality: Television and Proper Distance." *International Journal of Cultural Studies* 14, no. 4 (1 July 2011): 383–400.

———. *The Poetics of Digital Media*. Cambridge: Polity, 2018.

Hand, Sean, ed. *The Levinas Reader*. Oxford: Wiley, 2001.

Heidegger, Martin. *Being and Time*. San Francisco, CA: Harper and Row, 1962 [1927].

Highmore, Ben. *Ordinary Lives: Studies in the Everyday*. London: Routledge, 2010.

Ingold, Tim. *The Perception of the Environment: Essays on Livelihood, Dwelling and Skill*. London: Routledge, 2000.

Lagerkvist, Amanda. "Existential Media: Toward a Theorization of Digital Thrownness." *New Media & Society* 19, no. 1 (1 January 2017): 96–110.

Levinas, Emmanuel. *Otherwise Than Being or Beyond Essence*. New York: Springer, 2013 [1974].

———. *Proper Names*. London: Athlone Press, 1996 [1976].

Moores, Shaun. "Digital Orientations: 'Ways of the Hand' and Practical Knowing in Media Uses and Other Manual Activities." *Mobile Media & Communication* 2, no. 2 (1 May 2014): 196–208.

Pedwell, Carolyn. "Mediated Habits: Images, Networked Affect and Social Change." *Subjectivity* 10, no. 2 (1 July 2017): 147–69.

Pinchevski, Amit. *By Way of Interruption: Levinas and the Ethics of Communication*. Pittsburgh, PA: Duquesne University Press, 2005.

———. "Emmanuel Levinas: Contact and Interruption." In *Philosophical Profiles in the Theory of Communication*, ed. Jason Hannan. New York: Peter Lang, 2012, 343–66.

Sartre, Jean-Paul. *Being and Nothingness*. London: Simon and Schuster, 1992.

Slaby, Jan, and Gerhard Thonhauser. "Heidegger and the Affective (Un)Grounding of Politics." In *Heidegger on Affect*, ed. Christos Hadjioannou. Basingstoke: Palgrave Macmillan, 2019.

Sontag, Susan. *Regarding the Pain of Others*. London: Penguin, 2013.

Stewart, Kathleen. *Ordinary Affects*. Durham, NC: Duke University Press, 2007.

Withy, Katherine. "Owned Emotions: Affective Excellence in Heidegger and Aristotle." In *Heidegger, Authenticity, and the Self: Themes from Division Two of Being and Time*, ed. Denis McManus. Abingdon: Routledge, 2015, 21–36.

Chapter Two

Screen Time and the Young Brain

A Contemporary Moral Panic?

Ingrid Forsler and Carina Guyard

In recent years, excessive screen time has been widely discussed in relation to children and young people. Parents are advised to limit the amount of time their kids spend using digital devices, such as smartphones, tablets, or computers, and there is a wide selection of apps that parents can use to monitor and manage their children's screen time. The arguments against spending too much time in front of different screens include fear of addiction, depression, and other medical conditions, but also an increasing focus on how excessive screen time and constant connection affect social and cognitive abilities. Compulsory engagement with online technologies is assumed to make individuals absentminded, easily distracted, and indifferent to whatever goes around in the physical environment (Blum-Ross and Livingstone, 2016; Kardefelt-Winther, 2017). The latter debate emanates from the assumption that people, especially children and adolescents, are unable to control their impulsive behavior in relation to digital media. This inability among young people to resist their smartphones, although it might have negative outcomes, has sometimes been referred to as a contemporary moral panic in the media debate (Malik, 2019; Orben, Etchells, and Przybylski, 2018; Therrien, 2018).

Moral panics often occur when a new media technology is introduced and the users of these new media show disapproved forms of behavior, such as passivity or aggression. Historically there have been panic campaigns over a wide range of so-called low culture; comic books, rock 'n' roll, video nasties, etc. that is believed to degenerate in particular the younger generation due to violent or vulgar content (Buckingham and Strandgaard Jensen, 2012; Carlsson, 2010; Drotner, 1999; Critcher, 2008). Increasingly though, the concerns in relation to new media technologies focus specifically on the *use* of the media rather than with any particular content. As Alicia Blum-Ross and Sonia Livingstone (2016; 2018) have shown, the term "screen time" indicates a homogenization of media activities that does not take different practices or modes of engagement into account, but only considers the amount of time spent online (see also Kardefelt-Winther, 2017, 14). Indeed, the evidence cited in reports about screen time is dominated by short-time, quantitative studies that do not consider the broader life contexts of children. Additionally, in line with previous moral panics, they tend to focus on risks rather than on the opportunities of new media practices (Blum-Ross and Livingstone, 2016, 13; Kardefelt-Winther, 2017, 10). Other more qualitative inclined studies on children and media use have responded to this imbalance by highlighting the particularities of different media forms and uses as well as how parents react differently to guidance regarding screen time depending on socioeconomic and cultural factors (e.g., Blum-Ross and Livingstone 2016, 2018; Boyd and Hargittai, 2013; Clark, 2013; Lee, 2013; Livingstone et al., 2015; Livingstone and Byrne, 2018). In this chapter, we wish to contribute to this body of research by questioning the dominant perspective on the impacts of excessive screen time on young people.

The dominant perspective in relation to the use of online technology are increasingly supported by simplified accounts of findings from neuroscience and psychology. In this cross-disciplinary field, excessive use of digital technology is suggested to change brain structures and cause inability to focus, pay attention, and think rationally. On the market there is a wide and growing selection of popular scientific self-help books with titles such as: *i-Minds—How Cell Phones, Computers, Gaming, and Social Media Are Changing Our Brains, Our Behavior, and the Evolution of Our Species* (Swingle, 2015), *Mind Change—How 21st Century Technology Is Leaving Its Mark on the Brain* (Greenfield, 2014), *The Shallows—What the Internet Is Doing to Our Brains* (Carr, 2010). The popular scientific concerns of online technologies capacity to damage brain structures has received a lot of attention in the mass media, spreading the idea of a "re-wired" or hijacked brain to an even wider public (Kardefelt-Winther, 2017, 23). This mixture of mass media debates and popular scientific books, finding support in neuroscientifically and psychologically based arguments of online media effects, will be referred to in the chapter as *popular neuropsychology discourse*.

The concern of damaged brain functions is mainly directed towards children and young people that are claimed to be particularly sensitive to a large amount of time in front of the screen since they have not yet fully developed brains (Crone and Konjin, 2018; Kardefelt-Winther, 2017; Livingstone and Byrne, 2018). According to Kirsten Drotner (1999), the construction of children and young people as a vulnerable group in need of protection is characteristic for media panics. It is based on the assumption that culture holds a civilizing potential and only by consuming the right kind of culture through the right kind of medium can children develop into responsible grown-ups:

> [C]ultural development and human development are aspects of one and the same process. Children's cultural edification is part of, indeed proof of, their social elevation. Therefore their cultural fare must be guarded, watched over and protected because its composition is vital for their mental growth. (Drotner, 1999, 611)

This chapter takes these concerns as a point of departure to discuss what the recent debate about young people's screen time and possible brain damages reveals about contemporary culture and society. Moral panics are ultimately about reconfirming cultural values and as such they offer "a unique possibility to gain insight into the ways in which the media invoke and serve to reflect fundamental social and cultural problematics" (Drotner, 1999, 597). To map the popular neuropsychology-based assumptions and imaginaries underpinning what we in the chapter will refer to as *a screen time panic*, we follow the analytical model of moral panics developed by Erich Goode and Nachman Ben-Yehuda (1994). Their model, which we explain below, defines five criteria with which a moral panic can be recognized and distinguished from other kinds of social problems.

PANIC! DEFINITIONS AND CONCEPTUAL DEVELOPMENT

Before exploring the screen time panic, the concept of a moral panic as such (as well as neighboring concepts such as media panic and technopanic) needs to be explained. In an often-cited definition, Stan Cohen describes moral panics as a

> condition, episode, person or group of persons emerges to become defined as a threat to societal values and interest; its nature is presented in a stylized and stereotypical fashion by the mass media; the moral barricades are manned by editors, bishops, politicians and other right-thinking people; socially accredited experts pronounce their diagnoses and solutions; ways of coping are evolved or (more often) resorted to; the condition then disappears, submerges or deteriorates and become more visible. (1972/2002,1)

While Cohen indeed acknowledges the role of the media in his seminal definition, other scholars writing specifically about the reactions evolving around media have preferred the label *media panic* developed by Drotner. In Cohen's definition the role of media was merely to make the problem known and condemned. Drotner, however, maintains that the concept "media panic" is more appropriate when "the media is both instigator and purveyor of the discussion" (1999, 596). In the literature exist different interpretations of what this means, with on the one hand those that apprehend this as "instances where the media themselves are the focus of panic rather than just the means by which it is spread" (Buckingham and Strandgaard Jensen, 2012, 414) and on the other hand those who read this as the media inventing the panic (Dalquist, 1998).

Another version of the media panic concept is *technopanic*, originally labeled by Alice Marwick (2008). She defines technopanics as follows: First, they focus on new media forms and practices such as hacking or file-sharing. Second, they pathologize young people's use of the media. Third, the anxiety that are invoked leads to attempts to modify or regulate this particular use. As Adam Thierer (2013) points out, technopanic as an intense response to the emergence or use of media technologies is a version of the moral panic concept but with clear emphasis on form over content:

> By extension, a "technopanic" is simply a moral panic centered on societal fears about a particular contemporary technology (or technological method or activity) instead of merely the content flowing over that technology or medium. (2013,7)

Even though the technopanic definition seems promising to understand contemporary panics over (especially) young people's excessive use of media technologies, the assumption that earlier media panics are mainly occupied with content is debatable. Historically, media forms have evoked as much panic or concern as the message that they mediate. This is famously illustrated by Walter Ong (1982/2002) and others through Socrates's complaints about writing and its impact on the human mind, as diminishing memory and prioritizing facts over deep knowledge. As Ong points out, these concerns are recurring in the debate about computers and calculators:

> Writing, Plato has Socrates say in the Phaedrus, is inhuman, pretending to establish outside the mind what in reality can be only in the mind. It is a thing, a manufactured product. The same of course is said of computers. Secondly, Plato's Socrates urges, writing destroys memory. Those who use writing will become forgetful, relying on an external resource for what they lack in internal resources. Writing weakens the mind. Today, parents and others fear that pocket calculators provide an external resource for what ought to be the internal resource of memorized multiplication tables. Calculators weaken the mind, relieve it of the work that keeps it strong. (1982/2002, 78)

The idea that new media forms harm the human brain and its cognitive abilities is in other words not a new one. What is new, however, is how this risk is discussed and who gets to speak. While older media panics have been instigated by representatives from academia and education (Ong, 1982/2002) or journalists and media workers (Drotner, 1999), the contemporary debate about brain development and screen time is dominated by accounts from the popular neuropsychology discourse.

This debate is further focused around the use of media technologies rather than on form more broadly, as in the term "technopanic" that also includes panic related to new technology that has nothing (or little) to do with actual use, such as radiation or cloning. Gerard Goggin (2006), for instance, argues that public fears about electromagnetic radiation from cell phones and transmission towers are early expressions of the panics related to the current cell phone culture. The technopanic concept might therefore be too broad and inclusive to capture the screen time panic discussed here. Media panic, on the other hand, seems too narrow and media centric to capture a phenomenon that—as discussed later in this chapter—to a great extent is focused on individual responsibility and self-governance rather than on media regulation. While also using literature about media panic and technopanic to discuss specific aspects of the screen time debate, this chapter draws mainly on the overarching concept of *moral panic.*

But why reiterate the old and much criticized concept moral panic at all? Angela McRobbie and Sarah Thornton (1995), for instance, suggest that the model of moral panic needs to be updated to remain relevant today. They are particularly critical of the morality aspect of the concept. Some scholars maintain that if the panic is not concerned with a *morally* deviant behavior it is not a "real" moral panic (e.g., Critcher, 2006, 3). McRobbie and Thornton, however, argue that maintaining a consensual social morality is becoming increasingly fruitless in a time when "nothing could be better for sales than a bit of controversy" (1995, 572). According to them, being radical and rebellious has actually become more desirable than to exhibit a morally acceptable behavior, not the least among youths. While recognizing that the idea of what moral behavior is might have changed, we argue that the term is still useful to understand the screen time panic. In the current risk society (Beck, 1992), risks "are often subject to processes of moralization as individuals are called upon to 'be responsible'" and thereby self-harm "is evidence of a failure to manage one's risk—indicating *irresponsibility*, a moral affront" (Wright Monod, 2017, 39). This means that the call to manage screen time and disciplining the brain could well be seen as a question of morality. We further argue that moral panic is useful as a "sensitizing concept" that might "suggest directions along which to look" at a certain phenomenon (Blumer, 1954, 7).

BRAINS AT RISK: FIVE CRITERIA

To distinguish a moral panic from other mediated debates on social problems, Erich Goode and Nachman Ben-Yehuda (1994) have developed an analytical model with five key criteria, including 1) a heightened level of *concern* over a new behavior that is assumed to be threatening to the rest of the society, 2) *hostility* against the group held responsible or engaging in the disapproved behavior, often portrayed as the enemy or as "folk devils," 3) *consensus* among the majority of the population that this is a serious threat, 4) that the concern is *disproportionate* to the nature of the threat, and 5) that moral panics are *volatile* and tend to disappear as quickly as they emerged. In the following we intend to discuss the screen time panic in relation to each of these criteria.

Concern

The most basic criteria of a moral panic is that it is occupied with a new kind of behavior or practice that is considered problematic and unwanted, not only on an individual level but also as a threat to the rest of the society. In other words, moral panics reflect a concern over a social problem, that is "manifested or measurable in concrete ways, through for example public opinion polls, media attention, proposed legislation, action groups, or social movement activity" (Goode and Ben-Yehuda, 1994, 157). Whereas some writers claim that a moral panic occurs when the threat "is felt to represent a crisis for that society" (Goode and Ben-Yehuda, 2006, 50), others settle for a behavior that is considered "a threat to societal values and interests" (Cohen, 1972/2002, 1). Whether it is assumed to be "a crisis" or merely "a threat," panics in one way or the other occur on a societal level.

The concerns of distracted, always connected youths cannot undoubtedly be said to constitute a moral panic in those terms. Considering the popular neuropsychology discourse inclination towards self-help advice, the threatening behavior becomes first and foremost an individual problem. The screen time panic does, however, correspond with the definition of a moral panic as addressing behavior that challenges norms and cultural values, not least those associated with modernity where "upbringing is seen as the locus of character formation" and consequently must be carried out to foster desired social values (Drotner,1999, 613). One such value in modern society is rationality, often juxtaposed against emotionality in moral panics historically (Drotner, 1999, 613). The screen time panic still holds rationality in high regards, but the threat is no longer violent or sexual content that might arouse strong emotions in the young mind, but rather how the constant interaction with screens provides "sensory and cognitive stimuli" that causes "hurried and distracted thinking, and superficial learning" (Carr, 2010, 116).

By being constantly distracted by their smartphones and other digital media devices, young people are believed to be incapable of engaging in time-consuming and cognitive demanding activities, such as reading longer texts or solving logical problems. This concern reveals a cultural ideal of how time ought to be spent. Since time is conceived as a "zero-sum commodity" (Kardefelt-Winther, 2017, 11), spending time on digital media will detract from other more "valuable" activities, such as reading books or exercising. However, within the popular neuropsychology discourse the concern goes beyond worries about wasting time. It also includes the possible effects of long-term exposure to digital media on the young brain, explained through ideas of the "plastic brain." Brain plasticity refers to "the capacity of the brain to modify itself in response to changes in its functioning or environment" (Pitts-Taylor, 2010, 636). This means that the human brain can be trained to perform better, just like other parts of the body improve with exercise, but also that the brain loses capacities that are not used. An adult brain is less plastic than a young brain, giving that the change in brain structures caused by stressful environments or repetitive behaviors early in life might become constant, or at least harder to get rid of, as we grow older (Carr, 2010; Greenfield, 2014; Klass, 2019). Constant calls for attention from digital devices are assumed to be stressful and might prevent the young brain from developing deep learning skills and, in the long run, strengthen this inability through lack of exercise. This raises concerns about the long-lasting effects of children's media use, not only as a "question of character formation" (Drotner, 1999, 615) but also as something that risks rewiring the structure of the young brain permanently:

> And, thanks again to the plasticity of the neuronal pathways, the more we use the Web, the more we train our brains to be distracted—to process information very quickly and very efficiently but without sustained attention. That helps explain why many of us find it hard to concentrate even when we're away from our computers. Our brains become inept at forgetting, inept at remembering. (Carr, 2010, 194)

The threat at the heart of the screen time panic is thus mainly concerned with individuals and their shrinking cognitive abilities, but this might in the long run challenge some of the central norms of the society, such as rationality and focused thinking. What will happen when today's young generation is in charge of organizing the society if they cannot be focused and think rationally?

Hostility

The second criteria of Goode and Ben-Yehuda's analytical model is that there must be an increased level of hostility towards the group engaged in the behavior in question. "That is, not only must the condition, phenomenon, or behavior be seen as threatening, but a clearly identifiable group in or segment of the society must be seen as *responsible* for the threat" (Goode and Ben-Yehuda, 2006, 52). Moreover, those upholding the dangerous behavior are seen as evil—"folk devils" in Cohen's (1972/2002, 66) terminology—and must be punished or rehabilitated to conform with established societal norms. The folk devils of past moral panics have been groups with little societal power, such as children and adolescents, working-class people, or immigrants. The proposed solutions to their deviant behavior have in turn included changes in legislation and policy, local action groups and programs as well as extended rights and commissions for authorities such as the police and social services (Cohen, 1972/2002; Critcher, 2006; Drotner, 1999, 614). In this "morality play of evil versus good" the deviant group is portrayed in a simplified and stereotypical way by the dominating voices in the media and public debate, leading to "dichotomization between 'them' and us'" (Goode and Ben-Yehuda, 1994,157).

The "folk devils" of the screen time panic are, similar to many previous panics, children and young people, but also to some degree parents that neglect limiting the screen time of their children. The "evil" behavior that ought to be punished is the inability to take the eyes off the screen and concentrate on more serious matters. However, the question of responsibility is ambiguous. Today we live in an era of "deep mediatization" (Couldry and Hepp, 2017) where not only interaction with other people becomes increasingly mediated, but also other aspects of life such as work, education, civil services, and consumption. There are also strong commercial interests in keeping people online, and tech companies strive to develop as many irresistible apps as possible (Alter, 2017). It is hence recognized that the contemporary society is increasingly saturated by digital media technologies and that these devices and software are developed to be addictive, but, nonetheless, the main responsibility is not put on governments or companies but on individual families. With few exceptions, such as the proposed bill to the US Congress to "prohibit social media companies from using practices that exploit human psychology or brain physiology" (Jeffrey, 2019), digital media overuse is expected to be dealt with on an individual level. One of the most common responses to this expectation is screen time guidelines for the family. For instance, the American Academy of Pediatrics (2018) has provided "health and safety tips" for parents since, as they argue, "in a world where children are 'growing up digital,' it's important to help them learn healthy concepts of digital use and citizenship." The tips include rather mundane recommendations, such as to "set limits," "create tech-free zones," and "encourage playtime" (n.p.).

This marks a slight shift compared to previous moral panics. Although children and young people are still targeted as the object of concern, screen time panic differs from earlier examples that often led up to some kind of intervention on a societal level, such as legislation or action programs to be carried out in education or other institutions. Instead of promoting such public initiatives, screen time panic relies on individual responsibility in line with a neoliberal ideology where "caring and welfare moral duties that were once assigned to civil society and governmental entities" are now expected to be solved through self-governance (Shamir, 2008, 10). This individualized rather than structural coping strategy can also explain the focus on family and private life. Not even the most severe opponents to constant availability and stressful distractions suggest abandoning digital technologies once and for all. As Trine Syvertsen has shown in her book about media resistance, it is commonly argued that "you can still use devices for work or school, but should try to avoid digital entertainment and screens interfering with family time" (2017, 93). Tellingly, even high-level tech executives in Silicon Valley have begun to impose restrictions for screen time for their own children (Weller, 2018). This transfer of moral responsibility from institutions to individuals demands a shared understanding not only of who is responsible for taking action but also in what way and why—in other words, a consensus about the threat and what needs to be done.

Consensus

The third criteria in Goode and Ben-Yehudas (1994) panic model is that it must be a widespread agreement that the deviant behavior is in fact a threat. Chas Crichter (2006) identifies five groups with a capacity to define a deviant behavior as threatening: politicians, claims makers, law enforcement agencies, media, and the public. If all five groups unite over an issue as threatening, their power to define the problem is huge and oppositional voices will be swept aside. This is however seldom (or ever) the case. According to Critcher (2006, 10), to establish a panic over a deviant behavior it is sufficient that two of these groups enter into an alliance concerning the threat.

The deviant behavior discussed in this chapter is appointed by three of these groups, namely claims makers, the media, and the public. The key figures are the claims makers, or "socially accredited experts," with Cohen's (1972/2002, 1) terminology. In the panic of excessive screen time the main claims makers are, as argued above, writers and researchers within the field of neuropsychology. They are engaged as experts, explaining the cognitive damages due to the increased use of digital technologies in an accessible language in popular scientific books. Their power in "the century of biology" (Rose, 2013, 8) are based on their knowledge of the most objective and trustworthy source of information—the human brain. Today there is "an unparalleled truth discourse" about the brain and the human body (Rose, 2013, 7). As Victoria Pitts-Taylor writes, "the ability to know key truths about the self and the social are dependent upon developments in neuroscience" (2010, 635).

These experts also offer a form of life coaching, an industry that has become a huge business (Cederström and Spicer, 2015). The main idea in all forms of coaching is self-governance, corresponding to the logics of *responsibilization* (Shamir, 2008) discussed above, where "one of the central motifs running through most life-coaching interventions is that you must take responsibility for your own life and your own sense of wellbeing" (Cederström and Spicer, 2015, 13). This is evident in popular neuropsychology as well, which contains numerous suggestions about brain exercises in order to become more focused and more effective (Griffey, 2018; Klingberg, 2009). The market for these kinds of self-help books and courses seems insatiable:

> Bookshelves groan under the weight of popular science discussing this new knowledge of our biology, and speculating about the implications for our capacity to understand and control everything from our cognitive capacities to ageing and death. (Rose, 2013, 7)

Media professionals, such as science journalists, are the second powerful group in defining this particular moral panic. After all, much of what people know about the brain comes from "the press and from the experts in the self-help market who attempt to reach lay audiences through various kinds of media" (Pitts-Taylor, 2010, 641). Media representations of simplified findings from new brain science are ubiquitous, as well as discussions about possible dangers with an overuse of media technologies (Pitts-Taylor, 2010).

Closely connected to the media is the third group with power to establish the screen time panic as a real threat, namely the public. As Blum-Ross and Livingstone point out, parents are "inundated with guidance about screen time" not only by experts in popular media but also by other parents "through informal chats at the school gates" (2016, 12). The rise of "self-help tips" on how to manage this threat can also be detected in numerous social platforms, often in the form of testimonials "where individuals share experiences with media detox and abstention in the public sphere" (Syvertsen, 2017, 90). While these testimonials and advices often are directed to adults, the suggested strategies also include children and young people. For instance, Common Sense Media,[1] which is a nonprofit organization providing research and guidelines about media use for parents, is also used as a social network where parents share their own experiences and give advice to other parents. There are also many self-help books picturing how families struggle with ubiquitous media eroding family ties, such as Susan Mausharts bestselling book *The*

Winter of Our Disconnect (2011). These testimonials and various strategies to take control over a presumed unhealthy digital use confirm a consensus that spending too much time online is harmful, especially when it comes to children and young people.

Disproportionality

The fourth criteria of a moral panic according to Goode and Ben-Yehuda (1994) has to do with disproportionality, that is that the public concern over the behavior and the problem it poses exaggerates the scale of the problem "in reality." The criteria of disproportionality have been disputed by other writers since it assumes that there is an objective reality with which irrational panics can be compared (Buckingham and Strandgaard Jensen, 2012; Wright Monod, 2017). We agree with this critique, not least regarding the conceptualization of moral panics as social constructions where "social problems do not exist objectively; they are constructed by the human mind, called into being or constituted by the definitional process" (Goode and Ben-Yehuda, 1994, 151). However, the aspect of disproportionality can be used to discuss an imbalance of how the problem of screen time is approached and how objectivity is constructed discursively in media representations.

To begin with, it must be established that research on the relation between screen time and brain effects or well-being has not yet given any clear answers (Kardefelt-Winther, 2017). As previously discussed, voices from popular neuropsychology are convinced of the negative consequences of screen time for well-being and cognitive development but on a closer scrutiny, most of the referred studies assume a very cautious position. In an article about "digital dementia" due to spending too many hours on a daily basis online, MD Larry Dossey (2014) first states that previous research has shown that "eight hours or more of daily internet involvement with video games is correlated with brain shrinkage and damage in adolescents" (2014, 72) but he also admits that it is still unclear if this fact is true for all engagement with internet content. In other words, we do not yet know if "heavy exposure to online educational material [is] as damaging to young brains as playing video games" (Dossey, 2014, 72). The psychologists Eveline Crone and Elly Konjin (2018) likewise conclude that while adolescents' have underdeveloped neural systems they *might* be more sensitive to media exposure, a "critical question that remains largely unanswered is how adolescents' abundant media use may impact them developmentally in terms of structural brain development, functional brain development, and related behavior" (n.p.). Even representatives from within the field of cognitive neurosciences, such as Torkel Klingberg, argues *against* the otherwise widespread and easily bought worry that constant distractions from digital technologies make people unfocused and stressed:

> There is, fortunately, no research suggesting that exposure to mentally more demanding or challenging situations impairs our powers of concentration. Indeed, there is much that points to the contrary: it is in situations that push the boundaries of our abilities that we train our brains the most. (2009, 164)

So, is the threat of excessive screen time causing brain damages exaggerated? Of course, it is too early to tell. However, the cautious and sometimes contradictory statements about brain structures expressed in popular neuropsychology discourse suggest that we are maybe dealing with a moral panic. There is still not sufficient knowledge about how the human brain functions. New findings occur almost on a daily basis (Pitts-Taylor, 2010; Rose, 2013).

Furthermore, the dominance of neuropsychological claims in the screen time panic also reflects a disproportionality in terms of voices and truth claims, where perspectives grounded in biology are seen as more trustworthy than perspectives from social sciences or humanities, despite the weak evidence of the former. Cultural anthropologist Joseph Dumit (2004) argues that this trust in neuroscience can be connected to the use of imaging techniques such as brain scans and computer visualizations, that although they are the result of a range of cultural and scientific negotiations and interpretations, are understood as more objective than other sources, reflecting "our current cultural semiotics that privileges machines over experts in terms of objectivity" (133).

Volatility: What Next?

The last criteria of moral panics in the framework used here is that they are *volatile,* erupting suddenly and fading away "nearly as suddenly" (Goode and Ben-Yehuda, 1994,158). One indication that the screen time panic has emerged rather suddenly are the results from Google trends that suggest that searches for the term "screen time" have quadrupled in the last ten years,[2] while searches on "screen time brains" show a similar growth curve only during the year of 2018.[3] If we accept the transient characteristic of moral panics, this trend will most likely be reversed in a few years' time and the interest in screen time might be replaced by some other issue regarding media use, behavior, and upbringing of children and adolescents.

The sudden fading away of a moral panic does not mean that concerns over a deviant behavior cannot be sustained over long periods of time, but rather that they become routinized or institutionalized. A moral panic that has faded away from the limelight might remain in institutionalized form as for example organizations, legislations, or informal interpersonal norms (Goode and Ben-Yehuda, 1994). The question is what will happen to the legacy of screen time panics: Will they be forgotten or picked up by institutions where the possible treats can inform education, legislation, and media and information policy? Or will they play into social norms and informal rules concerning tech-free situations (i.e., dinner) or zones (i.e., bedrooms)? Will neuropsychological perspectives continue to dominate the debate on media use, or will it be complemented with other theoretical and scientific traditions? Have we even seen the peak of this trend, or will it grow bigger before suddenly declining?

CONCLUSION

Based on the discussion above, we identify the debate on screen time and the young brain as a contemporary moral panic. Moreover, we argue that it is based on a neoliberal ideology where the responsibility to manage the supposed risks of extensive digital media use is put upon the individual. Similar to how personal health has become an individual responsibility, guaranteed through healthy eating and regular exercise, each person is now also expected to manage digital distractions by "training the brain" through a range of exercises and technologies. This requirement is connected to the neuropsychological idea of brain plasticity claiming that the brain can adapt to changing practices and environments, which means that a stimulating environment can help the brain to develop as desired while the "wrong" kind of stimuli and activities might rewire the brain in unwanted ways. The idea that brains can be improved fits well with the logics of responsibilization where neglecting to train the brain or offer a stimulating and (at least to some extent) tech-free environment for your children could be considered deeply immoral. As summoned by Pitts-Taylor "the development of plasticity discourse is highly compatible with the neoliberal pressures of self-care, personal responsibility, and constant flexibility" (2010, 640). This paradox between on the one hand indeterminacy, and on the other hand a prescribed approach of dealing with the possibilities offered by the plasticity of the brain can be compared to that identified by Drotner as central to modernity when she asks: "Now, how can we all develop an individuality that at the same time is socially determined? Obviously we cannot. Modernity is founded on a paradox of sameness and difference" (1999, 612).

Another paradox is that between weak and simplified arguments and claims of objectivity. As discussed above, the one-sided focus on neuropsychology in the screen time debate on the expense of more social perspectives indicates a shared imaginary of natural sciences and medicine as more objective and trustworthy than the perspectives on media effects offered by humanities and social sciences. The new connection biology/psychology also marks a shift in human ontology. Who we are is no longer defined in terms of inner psychological characteristics. Personhood "no longer concerns itself with the mind or the psyche. Mind is simply what the brain does" (Rose, 2007, 192). At the same time, popular neuropsychology often draws on metaphors and tropes rather than on "hard facts," and the arguments put forth often disclose a very romanticized view of off-line interaction. One example is Susan Greenfield's laments over too much screen time, where she writes that "every hour spent in front of a screen, however wonderful, or even beneficial, is an hour spent *not* holding someone's hand or breathing in sea air" (2014, 20–21). Besides evoking the idealistic pictures of love and freedom, it is not an argument that holds for closer scrutiny.

Not only does it build on the Western middle-class premise that holding hands and breathing sea air is what would make most people happy, it also assumes that these activities cannot be combined with being online. In reality, it is of course possible to surf on the internet on the shore with your loved ones, while an hour off the screen could as well be spent staring into a wall completely alone. As researchers who study the contemporary trend with digital detox camps have shown, the pre-digital sociality is frequently idealized in technological discourses (Fish 2017; Sutton 2017). According to Sutton (2017) "digital technology has become the central culprit for the alienating aspects of modernity." This applies not least to the discourse on children and screen time where an "exposure to a natural environment or natural stimuli, may be seen as a useful and relevant intervention strategy to counteract the effect of exhausted cognitive capacities associated with overuse of smart technology" (Schilhab et al., 2018, 3). Tellingly, the American Academy of Pediatrics (2018) emphasize the importance of encouraging play and face-to-face interaction in the guidelines of screen time for parents.

But the idealization of the predigital sociality also has another dimension. We suggest that the threat attributed to the use of digital and smart technologies might have to do with temporality, more specifically the current shift from a "culture of speed" that relied on mechanical speed to a "culture of immediacy," brought on by ubiquitous media technologies. John Tomlinson defines this state as "a culture of instantaneity—a culture accustomed to rapid delivery, ubiquitous availability and the instant gratification of desires" (2007, 74). Previous generations have been socialized prior to the internet and therefore have a different conception of time. They perceive time as progressive, meaning that time-consuming, demanding tasks accomplished now might bring rewards later in life, and find the unwillingness among young people to do the same both incomprehensible and frightening. In other words, part of the concerns inherent in the screen time panic might be the notion that the young generations have no other conception of time than immediacy, the "here and now."

The occupation with *delayed gratification* in the screen time panic is perhaps best illustrated with the multiple references to the "marshmallow test." In the test that was performed at Stanford in the 1970s, four-year-old children were given the choice between having one marshmallow immediately or having two if they could wait alone, for up to 20 minutes, for a grown-up to enter the room. The study hypothesized that children capable of delayed gratification would be better off as grown-ups, based on their ability to postpone a reward (Shoda, Mischel, and Peake, 1990). The idea that moral capabilities are coupled with cognitive abilities and future success fits the screen time panic framework perfectly, and the marshmallow test has been used as a reference in a multitude of articles and studies, such as in the research project *The Digital Marshmallow Test*[4] where a specific app is used to map and understand impulsivity control among individuals, or in literature describing "the ability to resist a blinking inbox or a buzzing phone" during school work as "the new marshmallow test" (Murphy Paul, 2013).

From this perspective, fast digital technologies fostering a culture of immediacy could be seen not just as an inconvenience but as a real threat to society. The principle of delayed gratification is at the heart of the educational system as well as of the modern life itself. If the current digital environment encourages short time gratifications over patience and delayed rewards, the whole system will fail. The folk devils in this morality play are the children who keep getting distracted by digital devices and those parents too permissive to help them out of this addiction. The good ones, in turn, are those disciplined enough to limit their screen time or helping their children to do so, preferably by guarding their own media habits and setting a good example. Morality has become a matter of disciplining the brain and managing time.

NOTES

1. "Reviews for what your kids want to watch (before they watch it)," common sense media, accessed July 14, 2020, https://www.commonsensemedia.org/.
2. "Google trends," accessed July 14, 2020, https://rb.gy/cvbuae.
3. "Google trends," accessed July 14, 2020, https://rb.gy/inzr3w.
4. "The Digital Marsmallow Test," Northwell Education, accessed July 14, 2020, http://digitalmarshmallow.org.

REFERENCES

Alter, A. (2017). *Irresistible: The rise of addictive technology and the business of keeping us hooked.* New York: Penguin Press.

American Academy of Pediatrics. (2018, January 5). "Children and media tips from the American Academy of Pediatrics." Retrieved from https://www.aap.org/en-us/

about-the-aap/aap-press-room/news-features-and-safety-tips/Pages/Children-and-Media-Tips.aspx.

Beck, U. (1992). *Risk society: Towards a new modernity*. London: Sage.

Blum-Ross, A. and Livingstone, S. (2016). "Families and screen time: Current advice and emerging research." Media Policy Brief 17. London: Media Policy Project, London School of Economics and Political Science.

———. (2018). "The Trouble with 'Screen Time' Rules," 179–87, in Mascheroni, G., Ponte, C., and Jorge, A (eds.), *Digital parenting. The challenges for families in the digital age*. Göteborg: Nordicom.

Blumer, H. (1954). "What is wrong with social theory." *American Sociological Review*, 19(1), 3–10. https://doi.org/10.2307/2088165

Boyd, D., and Hargittai, E. (2013). "Connected and concerned: Variation in parents' online safety concerns." *Policy & Internet*, 5(3), 245–69. https://doi.org/10.1002/1944-2866.POI332.

Buckingham, D., and Strandgaard Jensen, H. (2012). "Beyond "media panics": Reconceptualising public debates about children and media." *Journal of Children and Media*, 6(4), 413–29.

Carlsson, U. (2010). "Young people in the digital media culture. Global and Nordic perspectives. An introduction." In U. Carlsson (Ed.), *Children and youth in the digital media culture: From a nordic horizon*, 9–22. Göteborg: Nordicom.

Carr, N. (2010). *The shallows: What the internet is doing to our brains* (1st ed.). New York: W.W. Norton.

Cederström, C., and Spicer, A. (2015). *The wellness syndrome*. Cambridge: Polity.

Clark, L. S. (2013). *Parent app—Understanding families in the digital age*. Oxford: Oxford University Press.

Cohen, S. (1972, 1987, 2002). *Folk devils and moral panics: The creation of the mods and rockers* (3rd ed.). New York: Routledge.

Couldry, N., and Hepp, A. (2017). *The mediated construction of reality*. Cambridge: Polity Press.

Critcher, C. (2006). *Critical readings: Moral panics and the media* (Issues in cultural and media studies). Maidenhead: Open University Press.

——— (2008). "Making waves: Historical aspects of public debates about children and mass media." In Drotner and Livingstone (eds.), *The international handbook of children, media and culture*. DOI: http://dx.doi.org.till.biblextern.sh.se/10.4135/9781848608436.n6.

Crone, E., and Konijn, E. (2018). "Media use and brain development during adolescence." *Nature Communications*, 9(1), 588.

Dalquist, U. (1998). *Större våld än nöden kräver? Mediedåldsdebatten i Sverige 1980–1995*. Diss. Lund: Univ.

Dossey, L. (2014). "FOMO, digital dementia, and our dangerous experiment." *Explore: The Journal of Science and Healing*, 10(2), 69–73.

Drotner, K. (1999). "Dangerous media? Panic discourses and dilemmas of modernity." *Paedagogica Historica*, 35(3), 593–619. https://doi.org/10.1080/0030923990350303.

Dumit, J. (2004). *Picturing personhood: Brain scans and diagnostic identity*. Princeton, NJ: Princeton University Press.

Fish, A. (2017). "Technology retreats and the politics of social media." *tripleC*, 15 (1), 355–69.
Goggin, G. (2006). *Cell phone culture: Mobile technology in everyday life*. London: Routledge.
Goode, E., and Ben-Yehuda, N. (2006). "Moral panics: An introduction." In Critcher, C. (ed.), *Critical readings: Moral panics and the media* (Issues in cultural and media studies). Maidenhead: Open University Press.
———. (1994). "Moral panics: Culture, politics, and social construction." *Annual Review of Sociology*, 20.1, 149–71.
Greenfield, S. (2014). *Mind change: How 21st century technology is leaving its mark on the brain*. London: Rider Books.
Griffey, H. (2018, October 14). "The lost art of concentration: Being distracted in a digital world." *The Guardian*. Retrieved from https://www.theguardian.com/lifeandstyle/2018/oct/14/the-lost-art-of-concentration-being-distracted-in-a-digital-world.
Jeffrey, C. (2019, July 31). "Congress wants to limit your social media time by law." TechSpot. Retrieved from https://www.techspot.com/news/81222-congress-wants-limit-social-media-time-law.html.
Kardefelt-Winther, D. (2017). "How does the time children spend using digital technology impact their mental well-being, social relationships and physical activity? An evidence-focused literature review." Unicef.
Klingberg, T. (2009). *The overflowing brain: Information overload and the limits of working memory*. Oxford: Oxford University Press.
Klass, P. (2019, November 4). "Screen use tied to children's brain development." *New York Times*. Retrieved from https://www.nytimes.com/2019/11/04/well/family/screen-use-tied-to-childrens-brain-development.html.
Lee, S. (2013). "Parental restrictive mediation of children's internet use: Effective for what and for whom?" *New Media & Society*, 15(4), 466–81. https://doi.org/10.1177/1461444812452412.
Livingstone, S., and Byrne, J. (2018). "Parenting in the digital age: The challenges of parental responsibility in comparative perspective." In G. Mascheroni, C. Ponte, and A. Jorge (eds.). *Digital parenting: The challenges for families in the digital age*. Göteborg: Nordicom.
Livingstone, S., Mascheroni, G., Dreier, M., Chaudron, S., and Lagae, K. (2015). *How parents of young children manage digital devices at home: The role of income, education and parental style*. London: EU Kids Online.
Malik, N. (2019, January 7). "Don't fall for the moral panic over children's screen time." *The Guardian*. Retrieved from https://www.theguardian.com/commentisfree/2019/jan/07/moral-panic-children-screen-time-phone-laptop.
Marwick, A. E. (2008). "To catch a predator? The MySpace moral panic." *First Monday*, 13(6). https://doi.org/10.5210/fm.v13i6.2152.
Maushart, S. (2011). *The winter of our disconnect: How three totally wired teenagers (and a mother who slept with her iPhone) pulled the plug on their technology and lived to tell the tale*. New York: Jeremy P. Tarcher/Penguin.
McRobbie, A., and Thornton, S. (1995). "Rethinking 'moral panic' for multi-mediated social worlds." *British Journal of Sociology*, 46(4), 559–74.

Murphy Paul, A. (2013). "The new marshmallow test: Resisting the temptations of the web." The Hechinger Report. Retrieved from https://hechingerreport.org/the-new-marshmallow-test-resisting-the-temptations-of-the-web/.

Ong, W. J. (1982/2002). *Orality and literacy: The technologizing of the word*. London: Routledge.

Orben, A., Etchells, P., and Przybylski, A. (2018, August 9). "Three problems with the debate around screen time." *The Guardian*. Retrieved from https://www.theguardian.com/science/head-quarters/2018/aug/09/three-problems-with-the-debate-around-screen-time.

Pitts-Taylor, V. (2010). "The plastic brain: Neoliberalism and the neuronal self." *Health*, 14(6), 635–52.

Rose, N. (2013). "The human sciences in a biological age." *Theory, Culture & Society*, 30(1), 3–34.

———. (2007). *The politics of life itself: Biomedicine, power, and subjectivity in the twenty-first century*. Princeton, NJ: Princeton University Press.

Schilhab, T. S. S., Stevenson, M. P., and Bentsen, P. (2018). "Contrasting screen-time and green-time: A case for using smart technology and nature to optimize learning processes." *Frontiers in Psychology*, 9/773. https://doi.org/10.3389/fpsyg.2018.00773.

Shamir, R. (2008). "The age of responsibilization: On market-embedded morality." *Economy and Society*, 37(1), 1–19. https://doi.org/10.1080/03085140701760833.

Shoda, Y., Mischel, W., and Peake, P. K. (1990). "Predicting adolescent cognitive and self-regulatory competencies from preschool delay of gratification: Identifying diagnostic conditions." *Developmental Psychology*, 26(6), 978–86. https://doi.org/10.1037/0012-1649.26.6.978.

Sutton, T. (2017). "Disconnect to reconnect: The food/technology metaphor in digital detoxing." *First Monday*, 22(6).

Swingle, M. (2015). *I-minds—How cell phones, computers, gaming, and social media are changing our brains, our behavior, and the evolution of our species*. Gabriola, BC: New Society Publishers.

Syvertsen, T. (2017). *Media resistance: Protest, dislike, abstention*. London: Palgrave Macmillan.

Therrien, A. (2018, June 21). "'Moral panic' over gaming disorder listing." BBC News. Retrieved from https://www.bbc.com/news/health-44560338.

Thierer, A. (2013). "Technopanics, threat inflation, and the danger of an information technology precautionary principle." *Minnesota Journal of Law, Science, and Technology* 14(1), 312–50.

Tomlinson, J. (2007). *The culture of speed: The coming of immediacy*. London: Sage.

Weller, C. (2018, February 18). "Silicon Valley parents are raising their kids tech-free—and it should be a red flag." *Business Insider*. Retrieved from https://www.businessinsider.com/silicon-valley-parents-raising-their-kids-tech-free-red-flag-2018-2.

Wright Monod, S. (2017). *Making sense of moral panics: A framework for research*. New York: Springer International Publishing.

Chapter Three

The Waves That Sweep Away

Older Internet Non- and Seldom-Users' Experiences of New Technologies and Digitalization

Magdalena Kania-Lundholm

The aim of this chapter is to empirically explore, understand, and discuss the digitalization of society as part of the shared and negotiated experience among older non- and seldom-users of networked technologies (ICT). More specifically, it focuses on how older people reflect upon social change brought by the so-called *waves of mediatization*, with a particular focus on digitalization (Couldry and Hepp, 2017). Digitalization is understood here as the third wave of mediatization that relates to the computer, internet, and mobile phone. Of particular interest here is the internet, namely the contemporary infrastructure that links media devices with computers and large data centers. It is also an infrastructure that has undergone rather fast transformation from a "closed, publicly funded and publicly oriented network for specialist communication into a deeply commercialized, increasingly banal *space for the conduct of social life itself*" (Couldry and Hepp, 2017, 50, emphasis in the original).

Mediation and the communicative organization of time are usually approached in terms of clocks, calendars, and timetables. However, being an older person is a specific position in the life course, which allows people to reflect back upon the social change prompted by technological development. It also offers a perspective on the lived experience of different *waves of mediatization*, including digitalization. This chapter departs from the idea that there is one universal and straightforward experience of time, speed, and technology but rather assumes *multiple temporal landscapes*, which come into play in different contexts and situations, including those when people (dis)engage with digital technologies (Wajcman, 2015). Consequently, this chapter asks: How does the digitalization of society and networked technologies, such as computers, smartphones, tablets, and social media, shape the subjective experiences of older internet seldom- and nonusers?

In the last decade, research on mediatization has grown expansively (Couldry, 2008; Deacon and Stanyer, 2014; Hepp et al., 2015). The concept itself does not refer directly to any specific single theory, but rather describes a general approach employed to critically analyze the "interrelation between changes in media and communication on the one hand and, changes in culture and society on the other" (Couldry and Hepp, 2013,197). Mediatization addresses one of the fundamental questions in social sciences and sociology in particular, namely that of social order and social change. In other words, it addresses questions such as "how the social order is established" and the relationship between social structure and individual agency. By focusing on the roles of the media as institutions, and the media as technologies and how these roles relate to one another, it offers a *processual* perspective on social change and transformation. Göran Bolin (2017) discusses three main approaches to mediatization, namely institutional, technological, and constructionist approaches. The institutional approach focuses mainly on the media as institutions and their relation to the society at large, the technological approach focuses mainly on the media and their technological affordances, while the constructionist approach has an ambition to understand how individual subjects *interact* with the media and "is more inclined to emphasize the interplay between media, individual agency and the formation of structural constraints" (Bolin, 2017, 21). This means that, despite some initial and anticipated forms of media use, individuals can relate differently to various media, depending on their past and present experiences. In this way, the constructionist approach to mediatization is sensitive to the context embeddedness of various media practices and offers a more dynamic and less causal approach to social change (Bolin, 2017, 21). This is also the approach employed in this chapter.

WAVES OF MEDIATIZATION AND NORMALIZATION OF THE DIGITAL

When it comes to capturing questions pertaining to social order and social change, Nick Couldry and Andreas Hepp (2017) offer a rather broad historical perspective and overview of media-technological innovations presented in the form of four major mediatization waves. The first two, namely *mechanization* and *electrification*, cover a long period of time, departing roughly from the mid-fifteenth century starting with the invention of the printing press, to the beginning of the nineteenth century and the widespread use of electricity that sparked inventions such as the telegraph, the camera, the telephone, and television to name just a few. However, they suggest that if we look back at the past century, especially departing from around the 1950s onward, we see particularly fast development and change when it comes to media-technological innovations. Consequently, one could argue that a person born in the 1930s or 1940s has experienced a rather wide variety of media in their lives, ranging from radio, television, audio tape, the computer and PC, ARPANET, the World Wide Web, social media, smartphones, and some forms of AI (artificial intelligence). The graph suggested by Couldry and Hepp ends in 2015 and shows the growing presence of smartphones and social media, but the authors mention *datafication* as possibly the fourth and most recent wave of mediatization with big data analytics, algorithms, AI, and the internet of things. There are at least two crucial implications and lessons that we learn from these four waves of mediatization.

First, and almost self-explanatory, is the issue of the fast pace of techno-social development and change. The topic of acceleration and more specifically the acceleration of late modernity became the foundation of the theory of *social acceleration* developed by Hartmut Rosa (2005). According to Rosa, the temporal dimensions of modernity are more fluid than the spatial ones, and that temporal dimension helps us better understand the transformation of contemporary modern society. He suggests that time is an inherent element and a principal dimension of communicative action and of how communication is part of the process of constructing the social world. According to Rosa, the acceleration of modern society consists of three dimensions, namely *technological acceleration*, *social change* and how it is perceived, and the *pace of life*, which includes an everyday experience of acceleration. All three dimensions contribute to what Judy Wajcman (2015) calls "the time pressure paradox," a contradictory subjective experience where on the one hand we have more time on our hands due to technological development, as the past two waves of digitalization and datafication imply, while on the other we feel busier than ever and have less time to carry out our tasks.

Second, the relationship between technological inventions and increasing mediatization results in the fusion of digitality, sociality, and networking, particularly present in the third and fourth waves of mediatization. This process has been triggered to a large extent by the dominant digital discourse (Fisher, 2010), which situates networked technology at the center of social transformation and emancipation. It refers to the close affinity between the late capitalist and post-Fordist society on the one hand and the legitimizing role of technology in it on the other. Consequently, this belief in technology results in the widespread and hegemonic assumption that technological solutions can be applied to nearly all, even the most complex social issues and problems. This techno-logic has also been previously described as the digital sublime (Mosco, 2017), internet-centrism, and solutionism (Morozov,

2013). Last but not least, these particular changes that characterize the latest wave(s) of mediatization have led to what Adi Kuntsman and Esperanza Miyake (2019) call *normalization of the digital*. The authors suggest that the recent body of scholarly work within digital studies has been shifting steadily towards "the algorithmic" area where politics, economy, culture, and society are increasingly dependent on "predictive flows of data" (Kuntsman and Miyake, 2019, 5). They point out that in the digital neoliberal era of proliferating "smart devices," obsession with metrics, tracking, and big data, contemporary life is technocentric and digital by default. This means that integrated systems organize our personal, work, and social lives. In other words, the latest waves of mediatization have reinforced what Luc Boltanski and Eve Chiapello (2005) call the "connexionist world" and José Van Dijck (2013) the "culture of connectivity." In this context, the relationship between humans and technology is conflated and technology often serves as a mirror of societal bias and injustice. In such a ubiquitous, hyperconnected digital age lies an imperative of connectivity that is often informed and supported by the predominant solutionist techno-logic.

CULTURE OF CONNECTIVITY AND THE PROBLEM OF DIGITAL (DIS)ENGAGEMENT

One aspect of the networked society is that the presence of digital technologies in the form of, for example, smartphones in everyday life, is increasingly taken for granted, to the point of rising concerns about digital addiction and information overload (Kissick, 2016; Haig, 2018). In result, questions of media resistance, digital withdrawal, and digital detox have become current topics on both public and scholarly agendas (Turkle, 2011; Light, 2014; Syvertsen and Enli, 2019). At the same time, the normalization of the digital has led to the proliferation of often policy-driven initiatives aiming to promote digital investments into both public and private sectors. Consequently, when digitalization and datafication become the dominant status quo that individuals have no or very little control over, other nondigital skills and services lose their significance. In such a context, various forms of ICT nonuse and involuntary digital disengagement are considered problematic. This issue has been best captured in the scholarly and public debate on the so-called *digital divide* that commonly distinguishes between the information "haves" and "have nots," users and nonusers alike (Millward, 2003; Morris et al., 2007). Even though more recent research has proved this picture to be much more nuanced than a simple use versus nonuse binary can capture (i.e., van Deursen and Helsper, 2015), it is often the case that certain groups that remain on the "have nots" end of the continuum are considered vulnerable, marginalized, and prone to further exclusion. This is the case, for instance, with older people who often remain on the receiving end of digitalization initiatives (Verdegem and Verhoest, 2009). To put it simply, the paradox of (the culture of) connectivity lies in the fact that on the one hand there are initiatives aiming to increase digital participation and promote digital solutions, while on the other hand there are increased concerns about information overload and digital addiction. In other words, too much of the digital is as bad as the lack of access to it. In both cases, however, the techno-solutionist logic dominates interventions that seek to solve these problems.

For instance, in the case of voluntary digital withdrawal, the tech market offers a variety of devices, services, and applications helping individuals to voluntarily "disconnect" in order to "reconnect," especially in the context of the workplace (Guyard and Kaun, 2018). In such cases, digital disengagement is considered necessary to maintain work productivity and effectiveness. A similar techno-logic dominates initiatives supporting digitally excluded groups. However, for marginalized, vulnerable groups such as older people, digital disengagement, whether voluntary or not, is often perceived as an aberration or exception, something to be fixed. Consequently, digital participation is promoted as a prerequisite for active involvement, social inclusion, and well-being in the digital society (Seifert and Rössel, 2019). In fact, initiatives and policies promoting the digitalization of services rest on the assumption that digital inclusion is synonymous with social inclusion. In both cases, however, the digital is taken for granted as the new normal that all social groups, regardless of the form of their digital (dis)engagement, need to accept. This assumption has been questioned recently in research and scholarly interventions coming from within both media and communication studies and social gerontology, respectively.

The norm of ubiquitous connectivity and digital engagement has been to some extent challenged by research on online disconnection, media refusal, and digital disengagement (Light, 2014; Karppi, 2018; Portwood-Stacer, 2012). Scholars also acknowledge that the continuous navigation between online and offline, use and nonuse of information technologies and devices has become a common, every day, banal, experience. For instance, Hesselberth (2017) understood disconnection as gesture/s to disconnect from particular digital devices, like a mobile phone or social media platforms such as Facebook as well as discourse on the "right to disconnect" involving practices of digital detox. Kuntsman and Miyake define digital disengagement as a *continuum* of practices that include online disconnection, withdrawal, opting out, leaving, nonuse, detox, unplugging, and more (2019, 6). The goal here is to challenge the one-dimensional understanding of digital refusal by suggesting that apart from various practices of disconnection, there are also different motivations to refuse or disconnect digital technologies. By approaching digital disengagement as a multidimensional continuum, Kuntsman and Miyake argue that we can imagine new possibilities of relations between technologies and freedoms, between engagement and digitality, sociality and refusal. They have also suggested using digital disengagement as a starting point to evaluate, both politically and empirically, each digital formation and policy in order to move away from "both normalized digitality and the pervasive nature of technological solutionism" (Kuntsman and Miyake, 2019,10). In this way, we can focus less on the interventions that aim to "fix" existing digital problems and instead pay more attention to actual users and their agency.

A similar, albeit differently formulated argument has been developed by scholars working within social gerontology. The ambition to challenge and question the norm of digital inclusion and participation comes from those who encourage us to rethink the role of agency in the process of coconstitution of aging and technology (Wanka and Gallistl, 2018; Peine and Neven, 2019). For instance, Alexander Peine and Louis Neven (2019) challenge the dominant interventionist logic of research into aging and technology that positions the internet and networked technologies as major solutions to the problems of aging. Instead, inspired by Bruno Latour's work and science and technology studies they suggest that aging and technology are not separate realms but are co-constituted and need to be studied together (Peine and Neven, 2019, 18). They also suggest that technology is often used in unexpected ways, something that emphasizes the agency of older people, especially when it comes to negotiating meaningful spaces for technology in their lives. In other words, older people can be seen as competent agents who actively participate and (dis)engage with technologies in multiple ways, rather than incompetent and frail laggards who mostly need support.

This chapter follows a similar path by departing from the idea of older non- and seldom-users as critical agents rather than a passive, marginalized group. The role of older ICT users and their critical capacities is a theme I have explored before (see Kania-Lundholm and Torres, 2018) where we argued for the importance of the "bottom-up" perspective of older people as media users.

The analysis presented here expands this notion by suggesting that the intersection of aging and technology use can offer further insights into how older people negotiate meaningful time and space for technology in their lives. It is also an approach that challenges the interventionist logic and instead suggests there is a dynamic relationship between technology (non)users and society. It is informed by Judy Wajcman's (2015) critique of existing contemporary social theories that are often, more or less explicitly, informed by a techno-deterministic, solutionist approach to the relationship between technology and society. Instead, she makes a case for the *social shaping of technology* approach, which challenges the dominant cultural imageries of technological development that directly link technological innovation and change with constant connectivity, efficiency, convenience, innovation, social inclusion, and progress. This means that both technological change in the form of "waves of mediatization" and technology use are unpredictable, open-ended, and often shaped by social, political, and economic factors. This approach supports the view that, even if in theory, technology is supposed to bring well-being and "good" to the society, the reality and practice are often messier than that. Wajcman acknowledges that we seldom have a chance to live outside of technologies, since they are "inextricably woven into the fabric of our lives" (Wajcman 2015, 2). This implies that, for instance, the experiences of acceleration and a sense of hurriedness are not universal and uniform but rather that various social groups experience them in different ways. Consequently, engaging with technology becomes less a question of "if" but rather "how" and "when." The remainder of this chapter addresses this particular question, namely how digital technologies, the digitalization of society and the subjective experiences of older internet seldom- and nonusers have been mutually shaped.

MATERIAL AND METHOD

The empirical data in this chapter consist of six focus group interviews conducted in Sweden in the autumn of 2017 with thirty older (sixty-five-plus) internet non- and seldom-users between the ages of sixty-eight and eighty-eight. Internet nonusers were defined as people who neither own a PC or a mobile phone with active internet connection nor possess skills how to use these devices, and seldom-users were defined as those who access the internet at least once a week and possess some skills and knowledge on how to access information online. A total of eighteen females and twelve males were recruited through local associations and clubs for the retirees. As mentioned earlier, the choice of method, namely focus group interviews, was informed by the constructionist approach to mediatization (Bolin, 2017). More specifically, this perspective focuses on the phenomenological dimension and deals with the ways individual subjects perceive media and social change from *within* their life-worlds as media users (Bolin, 2017). This perspective allows for a dynamic approach to social change and operates in terms of "dialectic dualities, and how media technologies and institutions are always embedded in social contexts" (Bolin, 2017, 21). Additionally, this method delivers the kind of material that focuses on socially negotiated rather than solely individual, subjective positions, opinions, or experiences. That is, the individual experiences and utterances are evaluated and renegotiated in relation to others' experiences in a focus group situation.

During the course of the interviews, which lasted about seventy to eighty minutes each, the study participants were first asked to say what comes to mind when they hear terms such as "digital technologies," "digital participation," "digital divide," "digital exclusion," "social media," and "paper-free society." First, as expected, it was rather difficult for the participants to relate to these abstract concepts. However, they were able to engage in a more vivid discussion when presented with open, broader questions about the digitalization of society. For instance: "Do you remember your first encounter with computers?," "What do you think about the expectation to become/remain digitally competent in today's society?," "What are the advantages and drawbacks of the networked, digital technologies?," "Can you imagine the world and your life without the internet in it?" In order to facilitate the discussion and obtain more spontaneous reactions, the participants were asked to comment on the headlines from the main Swedish dailies about older people, computers, and the internet. The interviews were conducted and transcribed into Swedish. For the purpose of this chapter, selected citations have been translated into English. The results of the analysis presented below focus more specifically on the issues pertaining to digitalization, social change, and time, informed by the informants' own experiences with the internet, computers, and other "new technologies."

COPING WITH DIGITALIZATION, COPING WITH CHANGE

The discussion about digitalization initiated in the focus groups triggered a reflection on how much the society had been changing since the participants were younger and how rapidly this change took place. The technological development was also something that participants discussed, along with social change in general. Although asked more specifically about the impact of digital, networked technologies on their lives, the participants were interested in pointing out the life course perspective and the experience it had given them. Thinking back and reflecting on one's life had given them the opportunity to acknowledge, and to some extent even appreciate, social change and different waves of mediatization that became part of their life experience. The participants' experiences intersected mainly with two waves of mediatization: electrification and digitalization. This means that when they were born, in the 1930s and 1940s, radio and cars were already there, and television sets were used in Swedish households from the mid-1950s. During their working life in the late 1970s up to the 2000s some of them became familiar with computers, word processors, mobile phones, and later on even social media and smartphones. When reflecting back on past times, some acknowledged the fast change and technological development. The excerpt below illustrates this.

> Daria: I mean . . . the change in society took place really fast . . . from horse and carriage . . . and me as a child traveling [this way] with my grandfather and today . . . I do not really keep up with technology of today.
>
> Anna: There are some things that come back when one thinks about the IT society . . . industrialization, very big things that took place back then with our parents and our grandparents.
>
> Elin: I think it is nice to be able to experience everything from the previous era, so to speak. (Group #6)

Elin, by pointing to the experience of the "previous era," points rather directly to the experience of the two waves of mediatization that have shaped her life. It could be argued that the opportunity to discuss digitalization in the group of peers triggered a reflection on the broader socio-technological change that the participants were part of. It allowed them to see the current development from a wider perspective, although initially they were specifically asked about digitalization. The participants acknowledged that the most recent development that has to do with digitalization is somehow special for them. One of the most common reflections shared across all focus groups was that development, especially when it comes to technological innovations, has been so fast that it has become rather difficult to keep up with. It has to do with the fact that the speed of development also requires fast adaptation, learning, and continuous acquisition of new skills—which is rather difficult to achieve for some people at older age. For instance, Kajsa, at seventy-seven years old and defined as a seldom-user, expressed it in the following way:

> Kajsa: The development has been so fast, so once one has learned something, something new comes along again and then one doesn't know at all how to proceed. (Group #3)

The experience of social acceleration as described by Kajsa is often connected to the need to keep up and stay updated, but at the same time it is linked to the sense of failure since the acquired knowledge and skills need to be quickly replaced by new ones. This is particularly the case with digital technologies since the knowledge and skills to operate a computer or a mobile phone are not as simple and definite compared with, for example, the skills needed to use a radio or a TV set. Such situations can trigger a sense of isolation, anxiety, or even stress. Some of the participants reported feeling stressed, both when they were expected to have certain computer skills but also when devices did not work properly. When asked whether having limited or no IT skills made them feel isolated, some responded that this was in fact the case:

> I: Do you ever experience feeling isolated in some way?
>
> Britta: When one doesn't know how to [use] computers then one feels very much left out [it is] very stressful.
>
> Lasse: For me this is not a problem since I have so much other stuff to do.
>
> Britta: It is not that I do not have other things to do, but sometimes I have to do something on the computer and when it doesn't work as it should it gets very stressful.
>
> (. . .)
>
> Pernilla: It all happens so fast and when it does not work then it takes forever, so one has to sit down for several hours. And if it still doesn't get sorted out (. . .) I feel this with email and SMS and all this type of pressure and fear missing something . . . even though I do not think it is something I should have read or answered, I still think that one gets the feeling of being a bit hunted. (Group #3)

This particular sense of failure when devices do not work as they should, combined with pressure to be responsive and available, can be experienced as stressful to people like Pernilla. Although a sense of hurriedness, the fast pace of life, and time pressure are rather common for people active in working life (Wajcman, 2015), it is not necessarily the case for older people who are retired. At the same time, they might be experiencing anxiety and stress for other reasons. These can include, for instance, lack of skills and knowledge regarding how to use a computer and/or difficulties adapting to and navigating the reality that most of social services are solely available online.

Manuel Menke (2017) introduces the concept of *media nostalgia* to describe the idea that individuals who have difficulties copying with social change, particularly pertaining to media change (such as difficulties adapting to the newly emerging media environment, including, for example, mediated communication and persistence of ICT) develop stress, become introverted, and well-being decreases. It is the manifestation of individuals coping with stress in media-saturated societies that is mentioned most often by those who perceive social change, including media ecology, as challenging. The concept of *media nostalgia* refers not to longing for media technologies as *objects* but "for the state of the world they represent (. . .). It is about being included and being able to participate back then and feeling overwhelmed by communication expectations or even (partially) unconnected and excluded today" (Menke, 2017,640). This type of nostalgia also serves as an indicator of peoples' perceptions of social change, for instance in the form of different waves of mediatization. Those, like many older seldom- and nonusers, who perceive change as challenging might refer to longing for a time before or without certain media. In fact, one gentleman expressed this rather vividly by saying that "everything before [digitalization] was idyllic."

I would argue that in the case of older internet seldom- and nonusers interviewed for this project, the sense of *media nostalgia* pertained to how older people were coping with the media-technological change that took place throughout their lives. The sense of media nostalgia as such is not necessarily limited to older people and can be experienced regardless of age, social status, skill, and familiarity with technology. However, in the case of the participants interviewed for this study, nostalgia especially referred to the past ten to fifteen years, as pointed out by several participants. For instance, the existing imperatives to use the internet and computers, especially for certain online services such as online banking, eHealth, or purchasing tickets are experienced as sources of stress and anxiety. One of the ways of expressing media nostalgia, particularly present in the analyzed material, was articulated as a resentment that sociality had been "lost."

RESENTMENT BECAUSE OF LOST SOCIALITY

The arrival of networked technologies, including mobile telephony and social media, has presented social actors with both opportunities and problems. For instance, being online, networked, and mobile requires the copresence of others but also modification and management of how people talk, make sense of situations, and interact with each other. As William Housley (2019) points out: "[d]igital devices can disrupt shared expectations within routine interactional flows, but they are also realigned and incorporated (. . .) in ways which render visible the character of these new mobile social technical affordances and the organizational and mundane features of everyday social life." The disruption of expectations relating to routine interaction flows and the continuously changing ways in which social actors interact with each other was experienced and discussed by the participants as particularly disturbing. Digital devices, such as computers and mobile phones, are targeted as responsible for the change in how people socialize these days. Excerpts from groups #6 and #2 illustrate this:

> Karin: I miss certain things . . . when I look back at what we discussed earlier about being able to purchase tickets, meet someone [in person].

Anna: Yes, exactly, talking [to people].

Karin: This is no longer [the case].

Anna: This and different alternatives [to the online digital].

Karin: Yes.

(. . .)

Ada: We will become hermits, every each of one sitting in their own little cabin.

Karin: Yes, exactly, when I was a child and [was] with my parents . . . before the TV came, we played cards.

Ada: The community.

Karin: Yes. (Group #6)

Bengt: If one sits in front of the [computer] screen all evening, as my wife says, what did people do before then?

Christoffer: They talked to each other.

(. . .)

Maria: People were more social, I think.

Christoffer: It's just that people socialized in different ways, it was more personal, since one had contacts, people, friends to meet. . . . Today one cannot do it, as one sits front of a computer instead. (Group #2)

The excerpts from the focus groups above point to at least two things: First, there is media nostalgia for the time "before" the computers, more specifically before the wave of digitalization. This period is often rather idealized or even mythologized as a "time of community and togetherness." Second, there is the mourning for certain forms of sociality and togetherness that are presumably lost and have been replaced by digital, distant, and isolating forms of computer-mediated interaction.

Some people who find digitalization challenging and who have been socialized to communicate mostly in a predigital manner, can experience online exchange and interaction as puzzling and difficult to understand. This was the case with several participants in my study. In other words, they experienced problems in navigating this new, somehow even hostile (due to fast development and lack of alternatives), media ecology. As either totally unfamiliar (nonusers) or acquainted only in a limited way (seldom-users) with these "new technologies," their focus was on the disruption of the "old ways" of socializing and togetherness. This is to some extent what Göran Bolin describes as *technostalgia*, namely the yearning for a kind of predigital connectness, preceding contemporary forms of social networking (2016, 256). He argues that technostalgia is a combination of mourning for dead media technologies in themselves combined with the disappearance of tangible materialities of some media, such as print newspapers, vinyl records, or mix tapes. Technostalgia is not only about the unique quality of the medium as such but also about the emotional state of the individual. As discussed above, older participants in this project expressed nostalgia for the time when technologies were more familiar to them. In their view, those "old technologies" did not interfere with the way people socialized before, whereas, according to them, the "new media" do interfere. Nevertheless, the participants in the study were well aware that digitalization is here to stay and they need to, albeit reluctantly, find the time and a place for computers in their lives in order to "keep up" with social change and remain independent.

IN-BETWEEN TIME OF COMPUTERS

Eight out of thirty participants in the sample could be defined as nonusers. These were Sören, Pernilla, Bosse, Ludde, Henrik, Olle, Daria, and Karin, all between seventy-one and eighty-eight years old. They all reported not having a computer, tablet, or a smartphone and no ICT skills. They also expressed no desire to acquire them as long as alternatives were available to, for example, the possibility to pay a bill with a paper invoice. However, these alternatives were also a source of concern since some companies charge an extra fee for offline services. The rest of the participants could be described as seldom-users. Some of them owned a desktop computer, others preferred to use a smartphone or a tablet. They also performed different activities while online, although most used the internet for searching for news (eleven participants) and/or writing emails (eight participants). What all of these seldom-users had in common was how they described their time on a computer. Particularly striking was the fact that they did not necessarily see the time spent on the computer as leisure or entertainment time but rather something that had to be done. This raises questions about the place of new technologies in their lives and the time they spent using them.

A generally occurring pattern was that participants perceived their online activities as among their daily duties, like paying bills or searching for a recipe. They also agreed that these activities often took up too much of their time; time that would otherwise be spent on other activities. Maria, seventy-four, complained that it took too much time to do things on a computer:

> Interviewer: Do you feel like you would like to continue learning more, that you will need to learn more [ICT skills]?

> Maria: Mmhm, I think I should learn a little bit more but then I think that it takes so much time too.
>
> I: Yes, please tell us more about it.
>
> Maria: Yes, well, I think I'm a bit more practical. I like to fix things at home and always have something going on. I think it takes so much time to do things on a computer. Suddenly, I look at the clock and oh, wow, I have been sitting one for hour.
>
> Bengt: Yes, the time goes fast.
>
> Maria: Yeah, it just rushes past. (Group #2)

Maria, who is married to Bengt (seventy-seven), admitted also that before her retirement she worked in a school kitchen and did not have any contact with computers at all. Only after retirement was she encouraged by her husband and children to take some IT lessons so she could keep up with the development and also remain independent. In her case, as with several other participants, the computer remained a tool to do things rather than being a source of pleasure. The time on the computer was experienced as "wasted" and "stolen" from other activities. This frequent construction of a computer as a tool rather than a leisure technology reflects results from previous studies. For instance, Christina Buse (2009) studied how computer technologies relate to experiences of work and leisure in retirement. Her research focused on the subjective meanings of activities such as work and leisure among older adults. Buse argues that for retired older internet users the boundaries between work, leisure, and retirement are contested and reconstructed in relation to technology use. As a result, many retirees are rather reluctant and hesitant to define computer technologies as leisure, which can suggest generational (old age) and class issues (the majority of her informants, as in my case, had a middle-class background). The results of my analysis confirm this. Consequently, the screen time for older seldom-users becomes at best the time "in-between" other activities. Some informants, like Ludde (eighty-six), said that they would rather "go and pick berries or play piano" than sit in front of the screen. Also, Lasse (eighty-one) from Group 3, cited earlier, mentioned that he had "other things to do." This suggests that some older people have an understanding of leisure that does not necessarily include screen time. In Sweden, the understanding of leisure time in relation to work dates back to the organization of the modern welfare society in the 1930s. It refers to the time outside of wage labor with both organized collective activities such as playing sports or hunting, and individual ones such as picking berries or mushrooms (Aléx and Hjelm, 2000). It is a modern phenomenon that most likely informs the understanding of "free time" among older people.

However, even if the boundaries between productive work time and free leisure time are reconstructed in old age and during retirement, there is a certain construction of time that informs dominant understandings of internet use.

The cultural imperative of speed, captured for instance in theories of social acceleration (Rosa, 2005), is often accompanied by the economic construction of time as an individual resource rather than a collective achievement. Consequently, time goes to waste if it is not optimized and spent "well." As Wajcman suggests in her recent work, "[E]fficiency—associated with individual discipline, superior management and increased productivity—is one of the most powerful organizing ideologies of Western culture" (2019, 59). The cult of productivity, efficiency, and time optimization constitutes a moral order where time discipline remains a moral standard and a virtue. This construction of time also devalues other forms of slow labor and time, such as caring, doing housework, or picking berries. In this context, the idea of computer time as wasted, as articulated by older seldom-users, speaks to discourses that attempts to resist the dominant construction of time as an economic resource. This is because the logics of capitalist temporality dominate Western society and are best exemplified by notions such as "time is money" (Sörensen and Wiksell, 2019, 253). In the context of the information economy, where data is often described as "the new oil," digital technologies and digitalization are perceived as dominant tools employed in order to improve efficiency, productivity, and optimization. In other words, time is something that needs to be "saved," otherwise there will be a decline in private social relations and devaluation of private life (Sörensen and Wiksell, 2019, 253). Consequently, the reluctant adaptation to this dominant temporality, turning it upside down, so to speak, among older seldom-users can be seen as accepting and valuing more the "unproductive," alternative time beyond or even without the internet and networked technologies.

CONCLUSION

The aim of this chapter has been to explore and discuss how one of the waves of mediatization, namely digitalization, becomes part of the shared and negotiated experience among older non- and seldom-users of networked technologies. The main purpose was to understand how networked technologies shape the subjective experiences of older internet seldom- and nonusers. Digital media and networked technologies are often associated with the dominant temporal regime of acceleration, efficiency, and urgency, informed by the idea of time as an economic, individual resource. In the culture of connectivity, the digital is often taken for granted and normalized, and the tech-logic of solutionism prevails. Older people are often stereotypically represented as "digital immigrants" (Prensky, 2001), information-poor laggards whose problems of computer and other technology nonuse need to be fixed. Consequently, they become one of the main targets of the programs and policies of digital inclusion and participation. However, as the theme of this volume illustrates, there are alternative temporalities that resist the dominant logic of acceleration and digitalization, and people can become entangled in a diversity of temporalities. The results of the analysis presented in this chapter show this is indeed the case with older people. The constructionist approach to social change employed here explored the subjective experience of internet seldom- and nonuse and offered a dynamic perspective, sensitive to the context embeddedness of their daily media practices. There are three main points that deserve attention here.

First, the particular position in the life course, and experience of at least two different waves of mediatization, namely electrification and digitalization, have given the participants a particular perspective with which they reflect on their media use. By perceiving digitalization as both fast social change and ascribed coercion, the participants have challenged the techno-optimistic, solutionist idea that technological development is synonymous with efficiency, inclusion, and convenience. Instead, they described their ways of coping with accelerated social reality and how digitalization has "swept them off their feet," as the title of this chapter suggests. Of particular interest is the decoupling of human sociality from the digital, networked imperative, articulated in the form of tech and/or media nostalgia (Bolin, 2016; Menke, 2017).

Second, the time of retirement from paid work has offered them space to renegotiate boundaries between productive and unproductive time for work and leisure. In this new context, the screen time spent on the computer often falls "in-between" other tasks, frequently considered as more important. This can be interpreted as part of the process of negotiation between various temporalities, but also points to their reluctance to adapt to the imperative of connectivity as something inherently desirable and good. One could assume that those of the informants who were acquainted with computers and other digital technologies earlier during their working life would strongly associate them with "work" and thus refuse to use them during retirement. That was, however, not necessarily the case. In fact, those who were to some extent familiar with how to use a computer and/or a smartphone seemed to be more interested in pursuing further learning as compared to those who did not use computers during their working life. This does not mean, however, that even those seldom-users did not critically reflect on the imperatives of connectivity and digitalization of society. Finally, I suggest that by reluctantly adapting to this new digital reality while at the same time acknowledging the consequences of digitalization, for example, how it has changed how people communicate and socialize, older non- and seldom-users exercise their agency. This is to say that the previous experience of life in the predigital era has allowed them to offer a critical reflection on the dominant constructions of time and networked technologies in the hyperconnected digital age. Of particular interest for future sociological explorations would be to understand, for instance, the impact and consequences of internet (non)use and digital (dis)engagement from a generational, comparative perspective that includes the latest wave of mediatization known as datafication (Couldry and Meijas, 2019). Additionally, further research could focus on the place of advanced technologies, such as the internet of things, artificial intelligence, and other networked technologies in the architecture of everyday life after retirement. More specifically, we could ask: How various temporal regimes come together when people of different ages make sense of their lives, and at what cost, in the postdigital era?

REFERENCES

Aléx, Peder, and Jonny Hjelm, (eds.). *Efter arbetet. Studier av svensk fritid*, Lund: Studentlitteratur, 2000.

Bolin, Göran. "Passion and nostalgia in generational media experiences," *European Journal of Cultural Studies* 19, no.3 (2016): 250–64.

———. *Media Generations. Experience, Identity and Mediatised Social Change*, London: Routledge, 2017.

Boltanski, Luc, and Eve Chiapello. *The New Spirit of Capitalism*, London: Verso, 2005.

Buse, Christina Eira. "When you retire, does everything become leisure? Information and communication technology use and the work/ leisure boundary in retirement," *New Media & Society*, 11, no.7 (2009): 1143–61.

Couldry, Nick. "Mediatization or Mediation? Alternative Understandings of the Emergent Space of Digital Storytelling," *New Media & Society*, 10, no.3 (2008): 373–91.

Couldry, Nick, and Andreas Hepp. *The Mediated Construction of Reality*, Cambridge: Polity Press, 2017.

———. "Conceptualizing Mediatization: Contexts, Traditions, Arguments," *Communication Theory*, 23, no.3 (2013): 191–202.

Couldry, Nick, and Ulisses Meijas. *The Cost of Connection. How Data is Colonizing Human Life and Appropriating it for Capitalism*, Stanford, CA: Stanford University Press, 2019.

Deacon, David, and James Stanyer. "Mediatization: Key concept or conceptual bandwagon?," *Media, Culture & Society*, 36, no.7 (2014): 1032–44.

Fisher, Eran. *Media and Capitalism in Digital Age: The Spirit of Networks*, London: Palgrave MacMillan, 2010.

Guyard, Carina, and Anne Kaun. "Workfulness: Governing the disobedient brain," *Journal of Cultural Economy*, 11, (2018): 535–48.

Haig, Matt. "Google wants to cure your phone addiction. How about that for irony?" *The Guardian*, May 10, 2018, accessed: November 5, 2018, https://www.theguardian.com/commentisfree/2018/may/10/google-phone-addiction-app.

Hepp, Andreas, Stig Hjarvard, and Knut Lundby. "Mediatization: Theorizing the interplay between media, culture and society," *Media, Culture & Society,* 37, no.2 (2015): 314–24.

Hesselberth, Pepita. "Discourses on disconnectivity and the right to disconnect," *New Media & Society,* 20, no.5 (2017): 1994–2010.

Housley, William. "Reorienting sociology: Disruption and digital technology," *The Sociological Review Blog*, https://www.thesociologicalreview.com/reorienting-sociology-disruption-and-digital-technology/, March 29, 2019, accessed September 12, 2019.

Kania-Lundholm, Magdalena, and Sandra Torres. "Ideology, power and inclusion: Using the critical perspective to study how older ICT users make sense of digitization," *Media, Culture & Society,* 40, no.8 (2018): 1167–85.

Karppi, Tero. *Disconnect. Facebook's Affective Bonds*, Minneapolis: University of Minnesota Press, 2018.

Kissick, Dean. "Is the backlash against social media coming?" i-D, available at https://i-d.vice.com/en_us/article/is-the-backlash-against-social-media-coming, May 2, 2016, accessed June 17, 2019.

Kuntsman, Ari, and Esmeralda Miyake. "The paradox and continuum of digital disengagement: denaturalising digital sociality and technological connectivity," *Media, Culture & Society*, (vol xx) (2019): 1–13.

Light, Ben. *Disconnecting with Social Networking Sites.* London: Palgrave Macmillan, 2014.

Menke, Manuel. "Seeking comfort in past media: Modeling media nostalgia as a way of coping with media change," *International Journal of Communication*, 11 (2017): 626–47.

Millward, Peter. "The 'grey digital divide': Perception, exclusion and barriers of access to the internet for older people," *First Monday* 8, no.7 (2003). Retrieved from http://firstmonday.org/ojs/index.php/fm/article/view/1066/98614, 2018-05-21.

Morris, Anne, Joy Dean Goodman, and Helena Branding. "The 'grey digital divide': Perception, exclusion and barriers," *Univ. Access in the Information Society*, no. 6 (2007): 43–57.

Morozov, Evgeny. *To Save Everything, Click Here: Technology, Solutionism and the Urge to Fix Problems That Don't Exist*, London; New York: Penguin, 2013.

Mosco, Vincent. *Becoming Digital. Toward a Post-Internet Society*, London: Emerald Publishing Limited, 2017.

Peine, Alexander, and Louis Neven. "From intervention to co-constitution: New directions in theorizing about aging and technology," *The Gerontologist*, 59, no. 1 (2019): 15–21.

Portwood-Stacer, Laura. "Media refusal and conspicuous non-consumption: The performative and political dimensions of Facebook abstention," *New Media & Society* no.5 (2012): 1–17.

Prensky, Mark. "Digital natives, digital immigrants," *On the Horizon*, 9, no.5 (2001): 1–6.

Rosa, Hartmut. *Social Acceleration. A New Theory of Modernity*, New York: New York University Press, 2005 [2013].

Seifert, Alexander, and Jörgen Rössel. "Digital participation," in D. Gu and M. E. Dupre (eds.). *Encyclopedia of Gerontology and Population Aging*, Basel, Switzerland: Springer Nature, 2019

Sörensen, Majken Jul, and Kristin Wiksell. "Constructive resistance to the dominant capitalist temporality," *Sociologisk Forskning*, 56, nos.3–4 (2019): 253–74.

Syvertsen, Trine, and Gunn Enli. "Digital detox: Media resistance and the promise of authenticity," *Convergence: The International Journal of Research into New Media Technologies*, (2019): 1–15, https://doi.org/10.1177/1354856519847325.

Turkle, Sherry. *Alone Together: Why We Expect More from Technology and Less From Each Other*, New York: Basic Books, 2011.

van Deursen, Alexander, and Ellen Helsper. "A nuanced understanding of internet use and non-use amongst older adults," *European Journal of Communication*, 30, no.2 (2015): 171–187.

Van Dijck, José. *The Culture of Connectivity*, Oxford: Oxford University Press, 2013.

Verdegem, Pieter and Pascal Verhoest. "Profiling the non-users: Rethinking policy initiatives stimulating ICT acceptance," *Telecommunication Policy* 33, (2009): 642–52.

Wajcman, Judy. *Pressed for Time. The Acceleration of Life in Digital Capitalism*, Chicago: University of Chicago Press, 2015.

———. "Fitter, happier, more productive. Optimising time with technology," in V. King, B. Gerisch, and H. Rosa (eds.), *Lost in Perfection: The Impacts of Optimisation on Culture and Psyche*, London: Routledge, 2019.

Wanka, Anna, and Vera Gallistl. "Doing Age in a Digitized World—A material praxeology of aging with technology,"*Frontiers in Sociology*, 3, no.6 (2018): 1–16.

Chapter Four

Who Are the New Men in Grey?

Making Sense of Time, Time-Theft and Temporal Autonomy in the (Non)Use of Digital Media

Christian Schwarzenegger and Manuel Menke

In *Momo*, a novel written by the German author Michael Ende (1973), the Men in Grey are agents of the so-called Timesavings Bank. This mysterious agency makes the false promise that time saved in the present can be deposited in a bank and later, be returned with interest. But instead of being able to stock time for later use, the characters in the book become so dedicated to time saving that they increasingly forget to live in the present, in the moment. Doing so, life becomes increasingly sterile, as the time best suited for being saved is the time committed to social activities, leisure, art, creativity, or sleeping. Paradoxically, the more time people strive to save, the less they end up having, as the time they try to save is lost to them. But with all recreational moments, and with hours of doing *nothing in particular* gone, the rhythms of life become more hectic. Instead of helping to deposit the time saved, the Men in Grey consume all the time they helped to set free; they quite literally smoke it up in the form of cigars—the time is burned and blown away in the wind.

This chapter is about people's perceptions and reasoning about whether they have allowed new time-consuming Men in Grey to enter their lives. But instead of grey, these agents of time consumption come in bright colors, with flashing lights and through the swiping of shiny displays: the paradox of losing the time we are supposed to save also returns in this new guise—digital media and communication technologies in general may help us save time (Wajcman, 2015) in particular ways, for instance, we can easily overcome spatial and temporal constraints for communication and social (inter)action, allowing for a sense of accelerated communication and the impression that life indeed is accelerated. The same time-saving features of our technological devices can become a perceived source of stress (Reinecke et al., 2016;

Weinstein and Selman, 2016), as they prompt us to be available to our communicative connections and social relations at all times (Burchell, 2015; Thomas, Azmitia, and Whittaker, 2016); and our availability and activity in digital communication is partly monitored and made visible for other users of the same platform or the apps we use to connect. Rather than saving time for later use, (mobile) digital media and smart devices can hence tempt us to waste the time we gained using them. And as our devices and practices have increasingly become mobile as well, *in-between-times* (Görland, 2020), for example, times of waiting, periods of transfer, and times away from our mediated communication, tend to disappear, as those times are filled with the use of mobile media. In times when permanent availability and connectivity through digital media is considered the "new norm" (Klimmt et al., 2018; Vorderer et al., 2018; Vorderer and Kohring, 2013) and mobile digital communication has become pervasive, there is always something to do, something to explore, or something to miss out on. In this chapter, we approach the purported normalization of being permanently online and permanently connected through digital media and contrast it with personal accounts of how people make sense of time in a world imbued with ubiquitous and ever-present digital media communication and how they seek to maintain or regain discretionary autonomy over their time. Do people share a sense of acceleration through digital media communication as liberation from the constraints of time and the temporal situatedness of social life? Do media users share the experience that digital devices help them save time? Or is there a sense of a time-squeeze, as more social activities and options to act are perceived as available to them per temporal unit? How do users balance their sense of time and manage the time spent with their devices between the permanently available options to connect and a need to disconnect to have time for themselves, for work or to socialize offline? Traditionally, the time people spent with media was committed to the use of specific broadcasts, publications or other forms of media offerings—this time could thus be differentiated from the time without media (with the restriction that both radio and TV have also increasingly been used as background media, that is, a constant source of noise stimulation rather than focused use). However, the lines between media-time and time-without-media-use have become progressively blurred in the current "evertime of constant online connectivity" (Morrison and Gomez, 2014). Media users thus need to navigate more actively and responsibly between the multiple temporalities of media use and between on and off time with the media.

The research we present in this chapter is informed through our rereading and re-analysis of interview and media diary data we collected in a series of empirical studies conducted between 2014 and 2017. Our studies originally addressed different attempts by users to recalibrate, question, or adjust their relationship with digital media, and what role old and new media play in their life (Menke and Schwarzenegger, 2019) such as doing digital detox (Schwarzenegger, 2020; Schwarzenegger and Treré, 2018), abstention from media for religious reasons or engaging in mnemonic online communities in which they nostalgically criticize present media in contrast to their former media experiences (Menke and Kalinina, 2019). In the data we gathered, we found that in these different contexts and situations about which the informants of our studies reported, the question of time and the relationship between media and time as well as how temporal experiences are evocated, accelerated, or suppressed through digital media was a recurrent issue. The issue of time was foregrounded by them, even if it was not the focus of the initial interviews. Through the reanalysis of the original material, we have now aimed for a categorization of the temporal concepts, which were applied by the different groups of media users in the original studies, and which helped them to make sense of their digital media (non)use. We thereby found a complex interplay of various concepts of media-related time experiences, which make visible the layers of experiencing and valuating time in the everyday. To develop our argument, we first discuss the normalization of always connected and always online sentiments stabilizing the assumed ubiquitous omnipresence of digital media and how this shapes our understanding of offline time and time away from the media as well. In recent years, the topic of disconnection, media abstention, and media resistance found new prominence, highlighting the time without media as a test case for understanding the deep impact of digitalization and digital media pervasiveness on our sense of time, sociability, and understanding of culture and society at large. We then present the methodological take on our reanalysis and the original studies as well as the categorization of temporal understandings. We conclude with a discussion of the findings and what they suggest for future research.

NORMALIZING THE EVERTIME OF ONLINE CONNECTIVITY IN AN ACCELERATING SOCIETY

Thinking of change in society in terms of speed and an accelerated pace of life has a long tradition in social theory (Dodd and Wajcman, 2017). Further, discussing life as speeding up and having become faster than it used to be is a widespread perception, which not only finds academic resonance (Rosa, 2013; Tomlinson, 2007; Wajcman and Dodd, 2017) but also has gathered prominence in public debate. Imaginations of speed and acceleration have hence become somewhat "emblematic of our times" (Wajcman and Dodd, 2017).

It is also not new to think of this acceleration of society as being enabled or even brought about by the advent of new technologies, mostly technologies of communication and innovations by means of transportation and mobility. Acceleration through mobility and media have hence been discussed as a condition for the transformation of sociocultural practices on different scales, for example, the globalized expansion of social and cultural relations (Evans and Lundgren, 2016; Kern, 1983; Pentzold, 2018; Rantanen,1997, 2005; Urry, 2012). Giving a diagnosis, one which resembles a more recent digital media rhetoric of instant connection, timelessness, and immediacy, Kern noted as far back as 1983 that in an "age of intrusive electronic communication 'now' becomes an extended interval of time that could, indeed must, include events around the world" (314). With increased and accelerated communicative connectivity, the "world in reach" is extended both spatially and temporally, and simultaneously happening events from elsewhere become part of the here and now of mediatized individual lifeworlds (Schwarzenegger, 2017, 125–60). However, if this (meta)process or metanarrative of social acceleration and its link to new media and communication technologies is something that can be traced back decades and centuries (Carey, 2009), what is specific about the current period and its alleged evertime of permanent online connectivity?

The impression that "digitalization has spawned a new temporality" (Wajcman, 2015, 14) is quite pervasive. Also, earlier media have had an impact on temporal structures (Kaun, 2015) and imaginaries of time beyond the rather narrow reach of media use itself. Media have always had important functions for synchronizing and collectivizing the imagination of communities and societies (e.g., of national identities); time is constructed through mediation and media are temporalized (Kaun, Fornäs, and Ericson, 2016). But as mentioned above, times with media and times without media could have been more easily differentiated conceptually, empirically, and as a matter of experience. In an age of digital abundance with virtually all social domains affected by the presence of digital media communication and digital devices, this differentiation becomes increasingly problematic: "When we talk about media use today, we talk about activities that can no longer be understood as single units of action which can be narrowly contained in time, space, and usage" (Prommer, 2020, 299). Not only are media and communication technologies and the practices we perform with them and the operations they perform with and on us (e.g., tracking, tracing, personalizing) more deeply embedded in the fabric of everyday life than before. As Wajcman (2015, 10) describes, today "communication patterns and interpersonal sociability are much more mediated by and distributed across a whole range of multimodal devices." The most outstanding characteristic of this process is the extensive "possibility to connect permanently and everywhere" (Hepp and Hasebrink, 2018, 19). Through mobile digital media, knowledge, and entertainment, loved ones are accessible regardless of location. One can connect at any time and everywhere with the digital information repositories of the world, stream movies and music, play games, cultivate social relations, conduct business transactions, initiate romantic contacts, or purchase domestic inventory through online shopping. But with all those activities also comes the understanding that at the same time one always could also do something else as well. According to Vorderer and colleagues (2018), having the potential to be permanently online and permanently connected means that individuals increasingly have to divide their attention between the physically present situation and all the other digitally mediated events that are constantly happening at the same time, and in which they could partake as well. Prommer (2020) has described such and similar experiences as a polychronicity during simultaneity. She explains that "the ubiquity of mobile media and the possibility of spontaneous use has replaced linear consumption and transformed it into a polychrone time. Our sense of time has definitely been transformed into a polychronicity during simultaneity" (Prommer, 2020, 314). Neverla used the term of polychronous societies (Neverla and Trümper, 2020) to describe an abstract understanding of heterogeneous regimes of time that coexist parallel and interconnected, prompting the individual to navigate between different time options, time frames, and velocities of social life. The permanent potential of doing a multiplicity of things simultaneously bursts established "horizons of possibilities and expectations" (Rosa, 2017, 40), regarding norms and rules of usage of new media technologies and their affordances as well as their impact on social situations. People consequently have to adapt to new and transforming "social norms that evolve as devices are integrated into daily life" (Wajcman, 2015, 31), thereby transforming "the pace and scope of human interaction" (Wajcman, 2015, 10). Consequently, we do not only have the possibility to communicate faster and save time doing so, but to be permanently available and hence also responsive is increasingly something that is expected from us. As Wajcman (2015) puts it, "we inhabit a technologically suffused environment in which constant connectivity is the norm," or the "new normal" (Vorderer et al., 2018, 4). According to Wajcman (2019), this normalization goes hand in hand with a "driving cultural imperative for accelerated time handling to optimize productivity and minimize time wasting." In a way, it is the Silicon Valley where the New Men in Grey come from (Wajcman, 2019): we do now have means and applications to communicate faster and at any given time and place at our disposal, but likewise experience pressure and growing expectations concerning availability at every possible time and place (Mazmanian, Orlikowski, and Yates, 2013, 1341–51). Time spent away from our devices and abstaining from constant online connectivity becomes something that requires justification or must be explained and communicated about (Laurier, 2001). Morrison and Gomez (2014) hence used the notion of "evertime" to refer to the normalization of underlying expectations that everyone has their smart devices at hand and be connected to the online world anytime. These expectations can be implicit or explicit, so people might not even be aware of the time regime forced on them but still try to follow their sublime imperatives. Digital communication technologies can thus be seen at the core of a struggle between temporal autonomy and neoliberal cultural narratives of self-determination and the efficient use of time. Individual lifeworlds in the digital age can therefore be characterized by a tension between self-responsibility and social expectations to use time sensibly and efficiently and while doing so to stay connected and plugged in. The impact on the perception and sense making of time can be manifold. In current research, a lot of attention has been given to experiences of digital stress and negative consequences for individual well-being. In recent years research into practices of abstention, pushback, or resistance against the perceived overwhelming power and omnipresence of digital media can be found. When permanently online and connected becomes the new normal, disconnection and unplugging can become acts of defiance.

LIFEWORLDS IN OFFLINE-MODE? UNPLUGGING, DISCONNECTION, AND DETOX IN SEARCH OF TIME SOVEREIGNTY?

Zurstiege (2019, 28), in his recent book, *Tactics of De-connection and a Quest for Silence in the Digital Age*, emphasizes that current debates about living with digital media appear to invert previous assumptions of media theory. While in the past it was discussed under which conditions media users were willing to attribute human-like features to machines, it has increasingly become common to seek skills and competences in humans, which allow them to cope with the machines they have created. Humans now need to learn to tame their technologies' temptations, use their potentials in beneficial ways and resist their seductions. Hence, we can also witness an increased interest in and public calls for more media literacy. In an age of constant connection, Zurstiege (2019) argues the right of each and every one to be left alone, to keep their distance towards others, and to have space and time for their own discrete use is challenged and must actively be advocated for. Time without media and periods of the abstention from it are thereby not to be understood as "a special condition of some inspired individual" (Hesselberth, 2018), but rather an integral part of living with media. Resistance or abstention from the use of media technologies or certain services, in general or for a limited period, reflect the same processes of decision-making and reflections about media and their potential uses, as the decision to use do. Both outcomes are defined through people who position themselves in relation to the existence of media technologies and the normality of their use. Therefore, abstention from mediated (online) communication needs to be included as a legitimate stance towards media communication and nonusage has to be acknowledged and investigated as being one of the multilayered experiences of living in media saturated societies (Light and Cassidy, 2014; Kaun and Schwarzenegger, 2014). However, media and communication research archetypally is about studying, investigating, and understanding what people do with the media or what impact media have on them. Researching the nonuse of media, therefore, still is considered a somewhat puzzling and unfamiliar task for media scholars (Couldry, 2013). Communication scholars, with a few exceptions and until recently, have according to Woodstock (2014), largely ignored people who avoid media by choice and for considered reasons. In the past, media deniers and nonusers have mostly been studied through a clinical eye (Selwyn, 2003). Nonuse was considered a "problem," which should be "solved" (Selwyn, 2003). Following Hesselberth (2018), the nonuse of media technology was primarily framed as an issue of material or cognitive deficiency, belated adaptation, instigated by technophobia, ideological refusal, or nonacceptance. Even though people may give plausible reasons for nonuse, the body of research was quite judgemental towards nonusers. The unwillingness to spend time with (particular) media was rather treated as an "abnormality and deviation from the norm; a deficit to be overcome, a problem to be solved" (Hesselberth, 2018, 1996), than "a discursive move that entails more than simply not using something—it's a kind of conscious disavowal" (Portwood-Stacer, 2012, 1042). Meanwhile, media and communication scholarship has started to comprehend that studying reasons and practices of nonuse can help make visible the effects of exclusion when normalizing specific kinds of media usage—like permanently online, permanently connected—while alienating others (Kaun and Schwarzenegger, 2014). When media are mundane fellows and pervasive in all areas of social life they can be considered to be disappearing from our consciousness due to their omnipresence; looking at disconnections rather than connection exclusively enables us to develop a deeper understanding of what it means to live in an age when constant connectivity is considered normal:

> Listening to media resisters, an admitted minority, is important for the same reasons we value other minority groups and perspectives: They generate alternatives, they teach us about different ways of living, they remind us about

> the rapidity with which ways of communicating are changing and the inherent losses (as well as benefits) of those shifts. Resisters often articulate feeling out of step with the culture they see manifested in new communication technologies. (Woodstock, 2014, 1984)

Research has since identified disconnecting from digital media as motivated by the desire to reconnect with friends and ambitions to enhance the quality of relationships and spending more quality time with friends and family (Kuntsman and Miyake, 2019; Light and Cassidy, 2014). Disconnect to reconnect has since become a popular phrase in the digital detox movement and time spent offline is mythologised (Schwarzenegger and Treré, 2018) as more valuable than the time spent online, e.g., on social networking sites (Kuntsman and Miyake, 2015, 7–8). Disconnection from Facebook can, for example, be motivated in order to reconnect with people through other ways (Light and Cassidy, 2014; Karppi, 2018). When meeting friends in person or having dinner with the family is considered as more of a "quality time" than engaging with friends and family through digital media. We can also learn a lot about concepts of time wasted, time saved, or time well spent by observing how people make sense of disconnecting, detoxing, or not using. Disconnect to reconnect however does not only refer to reconnecting with yourself, family, friends, and nature, but perhaps, most importantly, to reconnect also with the digital media and platforms you have abstained from. Absence is considered temporary and not permanent. Unplugging is only to be done for a limited period and, in the words of Hartmut Rosa, with the goal of maintaining the functionality of the acceleration systems (Rosa, 2013, 50). Referring to them as oases of intentional deceleration (Rosa, 2013, 48), "oases of demediatization" (Hepp and Hasebrink, 2018), or "oases of de-digitalization" (Schwarzenegger, 2020) emphasizes how ambitions to slow down through disconnection cannot be seen as a detachment from acceleration society or permanently online lifestyles. Rather they are indicative of the permanent connectivity and accelerated life that demand a "space of self-reflection and controlled escape in order to remain manageable for us as human beings" (Hepp and Hasebrink, 2018, 18).

Ana Jorge (2019), in her analysis of user recounts of experiences of temporary disconnection on Instagram, found that disconnection is not a transformative action but rather restorative: they reinforce the social norms and expectations of being connected. Defiant practices of disconnection at the same time are increasingly commodified and integrated in the social media portfolio as representations of particular lifestyles (Jorge, 2019).

One such lifestyle element that has found prominence around aspirations of a more conscious and considerate use of media, which would ultimately also include more time away from media and to object the norm of constant connectivity is "slowness." Slowness as a concept and alternative has been around for a while, most prominently perhaps in the case of the slow food movement. Slow food as opposed to industrially processed alternatives is seen to be "from a richer ecological and ethical perspective, instead of a reductive industrial one focusing on speed, cost and efficiency" (Rauch, 2018, 14). Generally, deceleration as a response to perceived acceleration is not a new phenomenon either. Throughout the last decades, countermovements of slowing down accompanied nearly every major technological invention affecting the velocity of machines and processes (Wajcman, 2015, 174). Slowness in the context of media use is shifting the "usage toward slower mediated activities, often by temporarily or permanently reducing one's time spent with digital networks and devices" (Rauch, 2011, 1). Therefore, the richness in using media slowly is not only about using it less. It is about using media with the aim to gain more quality during times of media consumption, but also more quality of life in general. "Slowness" (Sharma, 2014) in contrast to acceleration and the permanently online, always available mind-set purportedly allows for more intense and genuine experiences, increased self-awareness, and more time for contemplation. Against this backdrop the sovereign choice to opt out and disconnect suddenly epitomizes the autonomy to make free decisions about how to spend time and to decelerate instead of being driven by the alleged accelerated affordances of life in modern late capitalist societies. In the next section of this chapter, we look at different types of temporal concepts, which were applied by the different groups of media users to make sense of their digital media (non)use.

FINDING TIMES WHILE SPENDING TIME TOGETHER: A CATEGORIZATION OF TIME CONCEPTS

The empirically informed categorization of time concepts we present in this chapter draws from our re-reading and re-analysis of data we originally had collected in a series of empirical studies conducted between 2014 and 2017. While pursuing different research interests and ends we used qualitative interviews and media diaries in all of them. In a way, the re-analysis of these studies, which originally were done in different contexts and independently, were brought together because of time. Time the authors spent together in a confined space: at the time the data for these different projects was collected, we were sharing an office at the University of Augsburg. Having conversations about what came up in the research we were doing at the time, we were struck by some apparent similarities in the data we both had collected.

The issue of time was foregrounded by our informants, both in interviews and their reflections in the media diaries. Informants were frequently using temporal metaphors or using narratives related to time to make sense of what they do with media, how they use them and what it means to abstain from usage. Further, the informants used references to time and experiences of temporality to make sense of the impact digital media have had on their lifeworlds. To approach our colloquially shared observations more systematically, we re-analyzed the data from these studies regarding (1) which concepts of time people employ to make sense of time in their (non)use of digital media, (2) through which practices and strategies they try to sustain/regain temporal autonomy when they feel that digital media "stole" time from them, (3) which past and contemporary imaginaries people draw from to envision a better life with(out) digital media and (4) identified how they normatively narrativize time.

The three studies we used for this secondary analysis related to reflections on time originally addressed a variety of attempts by users to recalibrate their relationship with digital media: people who participated in digital detox efforts ($n = 15$), people who temporarily but regularly abstain from media usage for religious ($n = 6$) reasons, that is, on Sabbath or during Lenten season, and lastly, people who participate in media nostalgic online communities ($n = 17$). Informants throughout the studies were diverse in terms of gender, age cohorts, professional and educational backgrounds, urban and rural dwelling areas as their perceived intensity of living in a permanently online, permanently connected world. We identified six different yet entangled concepts of time, which were characteristic for our informants' reasoning. These categories help us understand, through which arguments "offline" and "disconnection" have become symbols for slowness, tranquillity, revitalization and temporal autonomy as opposed to the rushed, accelerated pace of life in permanently digitally connected social worlds. They specifically allow to interrogate the role of communication technologies for how the accelerations and pressures of life are perceived as well as how the nonuse and use of certain technologies are appropriated to the lifeworld in order to reclaim time and temporal control from the media. All names used in the quotes from the empirical material are pseudonyms.

Overall, most of the categories we found in the data are related to somewhat critical positions towards how digital media enforce their time regimes on the user. We also could gradually identify some appreciation for the perceived situational autonomy over time gained by possibilities of asynchronic or global communication and fast information seeking. We call this category *Gained-time.*

> With fast internet connection, we think that we are doing a lot more than we are and that our productivity levels have increased because we are always connected. (Sarah, thirty-three years old)

These benefits of an accelerated communication were typically imagined against actual or imaginative experiences with how communication was more burdensome with dated technologies of the past. Zurstiege uses the metaphor of rowing to describe the relationship of digital media to the futures they inscribe and describe. According to him, the progress brought about by media technologies moves through time like a rower, facing towards the past and having his back turned to the future (Zurstiege, 2019, 26). The imaginations of what can be gained or is feared to be lost through technological innovations and new means of communication is modelled against the background of past experiences and media history. The category of Gained-time is closely linked to the concept category we call *Template-time*. Nostalgic memories of the past inspire imaginations oriented at "better times" in which media supposedly were less burdensome. They provide (typically highly normative) templates for better and more gregarious ways of living your life, than we are living it now.

> Kids today post everything on Facebook only so people give them likes. Back then we went outside without make-up, we played in the sand and climbed on trees. That is what I would rather want for my kids one day instead of all the new media. (Michaela, twenty-six years old)

The past is particularly used as a template for a better use of time without media, in nature with family and friends, making real experiences instead of virtual ones. However, these imaginations are rather used as a contrast and critique of the constantly connected today, and less a real longing for a restoration of the past. What also was striking in regard to the aforementioned category of Gained-time is that even those informants, who mentioned the potential of gaining time through accelerated communication, were well aware that this perception may be deceptive.

> But I think in some ways, it limits us as well and slows us down. Because you think you are multitasking, right? . . . So you are constantly pulled in different directions through your devices and so you're not as productive as you think you are. (Sarah, thirty-three years old)

Also, the category of Gained-time is not only applied to the benefits of using digital media technologies, but also to the contrary, time can be gained through abstention as well. Max, a forty-eight-year-old IT worker explains his detox experience:

> Once you have overcome the first hours, a kind of satisfaction occurs and you realize that it can also be done differently and that you have a lot more time for your kids and family. And then after one or two days, you realize that things still work.

A category that partly was surprising to us was what we called *Real-time*: spending time online and "in" digital media is often degraded as unreal and opposed to more authentic and direct experiences. Spending time offline in the "real" world was considered superior. A conceptual differentiation between real and virtual experiences online, which was quite prevalent also in research, especially in the first decade of the internet and has since been considered widely overcome, was still alive and well in the personal reasoning of our study participants—as it is in advertising campaigns for digital detox providers. We stick with the label of real-time, although we are aware that the use of real-time in this sense can be a bit confusing. In the context of digital media communication real-time is also used to describe completely different temporal experiences, namely instant and immediate communication, across distances, as if other persons were really there, regardless of their actual whereabouts. Real-time thus can be seen as the final frontier of accelerated communication with digital devices. The realness of temporal experience in the way we use it here, however, emerged as a strong in vivo code in all of the different studies. We thus think that it also beautifully illustrates some of the paradoxical temporal tensions in digital communication: the real-time of digital media is sometimes not deemed real enough. Real-time will often be related to nostalgic Template-time. Also, while the imagination that time spent without media would be somehow more conscious and valuable, informants were not so vocal about what to do in that valued time, but rather referred to common places and stereotypes of time with the family, sitting around campfires, singing songs and being real.

> The media are pulling us in pretty intensely and I told my mother: "Mum, I think you have spent time with us way more consciously than we do with our children today." (Amelie, thirty-seven years old)

Some informants feel somewhat besieged by the media, as one is permanently surrounded by them and the expectations, which they mediate.

> It is fair to say that my entire life, since a few years now, takes place in the media, I am permanently surrounded by them. Permanently connected. And increasingly I felt this might be too much. (Elke, forty-five years old)

Hence, the wish to get out of this state of being sieged and experiencing autonomous real time is nourished. In his book *Media Life* (2014), Mark Deuze uses the image of fish in the water to describe how people are living in the media instead of just with them. The category of Real-time suggests that informants consider themselves similarly submerged in a pool of media connectivity, but imagine that outside the water, something real is waiting for them, something specific, that still should be aspired and strived for. A special category of time the informants across the different studies were striving for was time for personal contemplation, self-awareness, and conscious reflection about themselves. The paths to the self were manifold. Religious rites and more earthly ambitions like detoxing during traveling or going to a spa could both be explained in terms of *Me-time.* Rosa (2013), in his reflections about the acceleration of society, explains that estrangement from the self can be one of the perceived consequences of perceived acceleration. This sense of being estranged and being disconnected from oneself could be found as characteristic for this category.

> And I decided like you know I want to spend some time to kind of getting in touch with who I am. (Sarah, thirty-three years old)

Disconnect to reconnect with who you really are and where you stand as a human.

> I was wondering how it would be to be without. What it would do with me as a human when I am not checking my phone or when I put the laptop aside—how would I spend my time. (Boris, fifty-four years old)

Abstaining from digital media allows focusing on oneself and Me-time helps to regain autonomy and control over your time and your life.

Regaining control from the media was a strong topic in the interviews as informants frequently felt overwhelmed, besieged by the media, and forced to follow their Siren calls. We describe this category as *Forced-time*: digital media demand participation due to their omnipresence and subsequent expectations by others. Users fear to disappoint others, miss out on important information or to ultimately be forgotten:

> And I said to myself, if I don't use it, I am stupid, because then I don't get the infos or maybe my friends forget about me. (Elsa, twenty-four years old).

The time people try to take away from the ever-presence is not really saved for later use, but rather our informants partially felt forced to catch up with what they missed during their absence:

> If I have been without my phone for three hours, then I will use it nonstop for half an hour afterwards (August, twenty-five years old); to compensate for the time I was offline (Peter, fifty-four years old).

The time informants experienced to be somewhat forced to reinvest in using digital technologies is sometimes experienced as our last category, namely *Lost-time.* Digital media are considered time-consuming with little benefit yet lots of distraction from more important things, but as they are distracting from more important tasks or focused work, the time spent with them often becomes invisible or eludes without being really noticed. We always have "so little time" Emma (twenty-five years old) explained, and as she is using her phone, she feels like time is flying by and she cannot even recall what she did other than "only looking at my phone."

CONCLUSION

While the scenarios at hand and the theoretical frameworks we have discussed so far are clearly linked to technologies and experiences of the digital age, the narrative structure of opposing time committed to media usage with more gregarious and socially acceptable tasks is not new, but a reoccurring topic in media and communication history. In their study on media rejection as a practice, Schwarzenegger and Kaun (2017) have discussed the movement called red dot, the international campaign against television. Schwarzenegger and Kaun (2017) had evaluated the entries in the campaigns "Pre-TV Generations Archive," a collection of entries documenting the decline of quality of living with the coming of the television. A further evaluation was based on the little white book of red dot. The arguments provided for time away from television echo arguments prominent in the 1970s in Germany, when the then chancellor Helmut Schmidt campaigned for a weekly "Day without TV," which was not intended as a practice directed against media, but rather as a means of positive family politics (Birkner, 2014).

In the chapter, we followed different narratives regarding the relation between media use and its potential for saving time. First, we revisited the narrative that technology will somehow "deliver us more time" (Wajcman, 2019, 1272), an imagination that is still being pushed by tech corporations today. Ideas that new media and communication technologies will allow for an accelerated communication, and communication freed from constraints of time and space would then become an ultimate tool to save time, represent a "long-standing but mistaken belief" (Wajcman 2019, 1272). However, the promise was not only that different kinds of social interaction and sociability could be performed faster but also more independently. New communication technologies repeatedly extended the "now" of communication, until it reached a state of always, ever and permanent. The possibility of being always available and permanently connected however, as we have discussed then, did not come without a toll, and affected social norms and expectations as devices were integrated in the fabric of everyday life.

In *Momo* it is the agents of the Timesavings Bank that come to town and trick the people into giving away their time, without gaining time. In the chapter, we have described new digital media and communication services as New Men in Grey, who tempt to spend more time with them, and their accelerated features of communication but ultimately provide a system of a comforting siege, making people actually spend more time with the devices than they are able to save. Now, the empirical impressions from the interviews suggest, saving time is rather associated with spending time away from the media, slow, conscious and in the real world. But this does not neutralize the Men in Grey. Quite the contrary, as appears to be the case, as we can see an increasing commodification of being offline, and a growing market of applications and services that are supposed to help users to regain autonomy over their time and save time consciously. And users, our research suggests, are willing to accept "foreign control" to help achieve regaining time from the media, that is, in the form of digital detox apps and other applications, helping you to control or become aware of your screen time spent with certain applications. For example, one of these apps vows to help "you avoid distracting websites. Block your own access to websites or mail servers for a pre-set length of time." This application, which monitors and regulates your online behavior, is called SelfControl. The Men in Grey do have a sense of irony. Tactics opposing the permanent connectivity end up being commodified by the same agents and technologies (Jorge, 2019), and thus the cycle of time saving and time wasting continues perpetually. Speaking of irony. Michael Ende, the author of *Momo* and the Men in Grey has also authored the fantasy novel *The Neverending Story* (1979). But that will be a different chapter.

REFERENCES

Birkner, Thomas. 2014. *Mann des gedruckten Wortes: Helmut Schmidt und die Medien.* Bremen: Ed. Temmen.

Burchell, Kenzie. 2015. "Tasking the Everyday: Where Mobile and Online Communication Take Time." *Mobile Media & Communication* 3 (1): 36–52. https://doi.org/10.1177/2050157914546711.

Carey, James W. 2009. "Space, Time, and Communications A Tribute to Harold Innis." In *Communication as Culture: Essays on Media and Society*, Rev. ed, 109–32. New York: Routledge.

Couldry, Nick. 2013. "Life without Media: Or, Why Mediacentrism Is Bad for You." In *Life without Media*, edited by Eva Comas, Joan Cuenca, and Klaus Zilles, 27–42. New York: Peter Lang.

Deuze, Mark. 2012. *Media Life*. Cambridge, UK ; Malden, MA: Polity Press.

Dodd, Nigel, and Judy Wajcman. 2017. "Simmel and Benjamin: Eearly Theorists of the Acceleration Society." In *The Sociology of Speed: Digital, Organizational, and Social Temporalities*, edited by Judy Wajcman and Nigel Dodd, First edition, 13–24. Oxford, United Kingdom: Oxford University Press.

Ende, Michael. 1973. *Momo: oder die seltsame Geschichte von den Zeit-Dieben und von dem Kind, das den Menschen die gestohlene Zeit zurückbrachte. Ein Märchen-Roman*. Stuttgart Wien: Thienemann.

———. 1979. *Die Unendliche Geschichte: Von A Bis Z*. Stuttgart: Thienemann.

Evans, Christine, and Lars Lundgren. 2016. "Geographies of Liveness: Time, Space, and Satellite Networks as Infrastructures of Live Television in the Our World Broadcast." *International Journal of Communication* 10: 5362–77.

Görland, Stephan O. 2020. "Really 'Dead Time'? Mobile Media Use and Time Perception in In-between Times." In *Mediated Time*, edited by Maren Hartmann, Elizabeth Prommer, Karin Deckner, and Stephan O. Görland, 321–40. Cham: Springer International Publishing. https://doi.org/10.1007/978-3-030-24950-2_16.

Hepp, Andreas, and Uwe Hasebrink. 2018. "Researching Transforming Communications in Times of Deep Mediatization: A Figurational Approach." In *Communicative Figurations*, edited by Andreas Hepp, Andreas Breiter, and Uwe Hasebrink, 15–48. Cham: Springer International Publishing. https://doi.org/10.1007/978-3-319-65584-0_2.

Hesselberth, Pepita. 2018. "Discourses on Disconnectivity and the Right to Disconnect." *New Media & Society* 20 (5): 1994–2010. https://doi.org/10.1177/1461444817711449.

Jorge, Ana. 2019. "Social Media, Interrupted: Users Recounting Temporary Disconnection on Instagram." *Social Media + Society* 5 (4): 205630511988169. https://doi.org/10.1177/2056305119881691.

Karppi, Tero. 2018. *Disconnect: Facebook's affective bonds*. Minneapolis: University of Minnesota Press.

Kaun, Anne. 2015. "Regimes of Time: Media Practices of the Dispossessed." *Time & Society* 24 (2): 221–43. https://doi.org/10.1177/0961463X15577276.

Kaun, Anne, and Christian Schwarzenegger. 2014. "'No Media, Less Life?' Online Disconnection in Mediatized Worlds." *First Monday* 19 (11).

Kaun, Anne, Johan Fornäs, and Staffan Ericson. 2016. "Media Times | Mediating Time—Temporalizing Media: Introduction." *International Journal of Communication* 10: 5362–77.

Kern, Stephen. 1983. *The Culture of Time and Space 1880-1918*. Cambridge, MA: Harvard University Press.

Klimmt, Christoph, Dorothée Hefner, Leonard Reinecke, Diana Rieger, and Peter Vorderer. 2018. "The Permanently Online and Permanently Connected Mind. Mapping the Cognitive Structures behind Mobile Internet Use." In *Permanently Online, Permanently Connected: Living and Communicating in a POPC World*, edited by Peter Vorderer, Dorothée Hefner, Leonard Reinecke, and Christoph Klimmt, 18–28. New York ; London: Routledge, Taylor & Francis Group.

Kuntsman, A., and E. Miyake. (2015). Paradoxes of Digital dis/engagement: Final Report. *Working Papers of the Communities & Culture Network, 6*.

Kuntsman, Adi, and Esperanza Miyake. 2019. "The Paradox and Continuum of Digital Disengagement: Denaturalising Digital Sociality and Technological Connectivity." *Media, Culture & Society* 41 (6): 901–13. https://doi.org/10.1177/0163443719853732.

Laurier, Eric. 2001. Why People Say Where They Are During Mobile Phone Calls. *Environment and Planning D: Society and Space, 19* (4): 485–504. doi:10.1068/d228t.

Light, Ben, and Elija Cassidy. 2014. "Strategies for the Suspension and Prevention of Connection: Rendering Disconnection as Socioeconomic Lubricant with Facebook." *New Media & Society* 16 (7): 1169–84. https://doi.org/10.1177/1461444814544002.

Mazmanian, Melissa, Wanda J. Orlikowski, and JoAnne Yates. 2013. "The Autonomy Paradox: The Implications of Mobile Email Devices for Knowledge Professionals." *Organization Science* 24 (5): 1337–57. https://doi.org/10.1287/orsc.1120.0806.

Menke, Manuel, and Ekaterina Kalinina. 2019. "Reclaiming Identity: GDR Lifeworld Memories in Digital Public Spheres." In *Communicating Memory & History*, edited by Nicole Maurantonio and David W. Park, 243–60. New York: Peter Lang. https://doi.org/10.3726/b14522.

Menke, Manuel, and Christian Schwarzenegger. 2019. "On the Relativity of Old and New Media: A Lifeworld Perspective." *Convergence: The International Journal of Research into New Media Technologies*, March, 135485651983448. https://doi.org/10.1177/1354856519834480.

Morrison, Stacey, and Ricardo Gomez. 2014. "Pushback. Epressions of Resistance to the 'Evertime' of Constant Online Connectivity." *First Monday* 19 (8).

Neverla, Irene, and Stefanie Trümper. 2020. "As Time Goes By: Tracking Polychronic Temporalities in Journalism and Mediated Memory." In *Mediated Time*, edited by Maren Hartmann, Elizabeth Prommer, Karin Deckner, and Stephan O. Görland, 219–37. Cham: Springer International Publishing. https://doi.org/10.1007/978-3030-24950-2_11.

Pentzold, Christian. 2018. "Between Moments and Millennia: Temporalising Mediatisation." *Media, Culture & Society* 40 (6): 927–37. https://doi.org/10.1177/0163443717752814.

Portwood-Stacer, Laura. 2012. "Media Refusal and Conspicuous Non-Consumption: The Performative and Political Dimensions of Facebook Abstention." *New Media & Society* 15 (7): 1041–57. https://doi.org/10.1177/1461444812465139.

Prommer, Elizabeth. 2020. "Polychronicity During Simultaneity: Mediated Time and Mobile Media." In *Mediated Time*, edited by Maren Hartmann, Elizabeth Prommer, Karin Deckner, and Stephan O. Görland, 299–319. Cham: Springer International Publishing. https://doi.org/10.1007/978-3-030-24950-2_15.

Rantanen, Terhi. 1997. "The Globalization of Electronic News in the 19th Century." *Media, Culture & Society* 19 (4): 605–20. https://doi.org/10.1177/016344397019004006.

———. 2005. *The Media and Globalization*. London; Thousand Oaks, CA: SAGE.

Rauch, Jennifer. 2011. "The Origin of Slow Media: Early Diffusion of a Cultural Innovation through Popular and Press Discourse, 2002–2010." *Transformations, 20*: 1–15.

———. 2018. *Slow Media: Why Slow Is Satisfying, Sustainable and Smart*. Oxford: Oxford University Press.

Reinecke, Leonard, Stefan Aufenanger, Manfred E. Beutel, Michael Dreier, Oliver Quiring, Birgit Stark, Klaus Wölfling, and Kai W. Müller. 2016. "Digital Stress over the Life Span: The Effects of Communication Load and Internet Multitasking on Perceived Stress and Psychological Health Impairments in a German Probability Sample." *Media Psychology* 20 (1): 90–115. https://doi.org/10.1080/15213269.2015.1121832.

Rosa, Hartmut. 2013. *Social Acceleration: A New Theory of Modernity*. New York: Columbia University Press.

———. 2017. "De-Synchronization, Dynamic Stabilization, Dispositional Squeeze: The Problem of Temporal Mismatch." In *The Sociology of Speed: Digital, Organizational, and Social Temporalities*, edited by Judy Wajcman and Nigel Dodd, First edition, 25–41. Oxford, UK: Oxford University Press.

Schwarzenegger, Christian. 2017. *Transnationale Lebenswelten: Europa als Kommunikationsraum*. Köln: Herbert von Halem Verlag.

———. 2020. "Mobil, vernetzt und digital—Kommunikationsräume und die Geographie der Lebenswelt." In *Räume digitaler Kommunikation: Lokalität—Imagination—Virtualisierung*, edited by Thomas Döbler, Christian Katzenbach, and Christian Pentzold. Köln: Halem.

Schwarzenegger, Christian, and Anne Kaun. 2017. "Switch Off! Media Rejection and Non-Usage of Media Technologies as a Resource for (Historical) Audience and Media Culture Research." ICA Annual Conference, San Diego.

Schwarzenegger, Christian, and Emiliano Treré. 2018. "Demystifying Disconnection: Deep Mediatization and Media Refusal." ECREA European Communication Conference, Lugano.

Selwyn, Neil. 2003. "Apart from Technology: Understanding People's Non-Use of Information and Communication Technologies in Everyday Life." *Technology in Society* 25 (1): 99–116. https://doi.org/10.1016/S0160-791X(02)00062-3.

Sharma, Sarah. 2014. *In the Meantime: Temporality and Cultural Politics*. Durham, NC: Duke University Press. http://public.eblib.com/choice/publicfullrecord.aspx?p=1632040.

Thomas, Virginia, Margarita Azmitia, and Steve Whittaker. 2016. "Unplugged: Exploring the Costs and Benefits of Constant Connection." *Computers in Human Behavior* 63 (October): 540–48. https://doi.org/10.1016/j.chb.2016.05.078.

Tomlinson, John. 2007. *The Culture of Speed: The Coming of Immediacy*. Los Angeles; London: SAGE.

Urry, John. 2012. *Mobilities*. Cambridge: Polity Press.

Vorderer, Peter. 2015. Der mediatisierte Lebenswandel: Permanently online, permanently connected. *Publizistik, 60* (3): 259–76. doi:10.1007/ s11616-015-0239-3.

Vorderer, Peter, Matthias Kohring. 2013. "Permanently Online: A Challenge for Media and Communication Research." *International Journal of Communication, 7*, 188–96.

Vorderer, Peter, Dorothée Hefner, Leonard Reinecke, and Christoph Klimmt. 2018. "Permanently Online and Permanently Connected: A New Paradigm in Communication Research?" In *Permanently Online, Permanently Connected: Living and Communicating in a POPC World*, edited by Peter Vorderer, Dorothée Hefner, Leonard Reinecke, and Christoph Klimmt, 3–9. New York; London: Routledge, Taylor & Francis Group.

Wajcman, Judy. 2015. *Pressed for Time: The Acceleration of Life in Digital Capitalism*. Chicago: The University of Chicago Press.

———. 2019. "How Silicon Valley Sets Time." *New Media & Society* 21 (6): 1272–89. https://doi.org/10.1177/1461444818820073.

Wajcman, Judy, and Nigel Dodd. 2017. "Introduction: The Powerful Are Fast, the Powerless Are Slow." In *The Sociology of Speed: Digital, Organizational, and Social Temporalities*, edited by Judy Wajcman and Nigel Dodd, First edition, 1–12. Oxford, United Kingdom: Oxford University Press.

Weinstein, Emily C., and Robert L. Selman. 2016. "Digital Stress: Adolescents' Personal Accounts." *New Media & Society* 18 (3): 391–409. https://doi.org/10.1177/1461444814543989.

Woodstock, Louise. 2014. "Media Resistance: Opportunities for Practice Theory and New Media Research." *International Journal of Communication* 8: 1983–2001.

Zurstiege, G. (2019). *Taktiken der Entnetzung: Die Sehnsucht nach Stille im digitalen Zeitalter*. Berlin: Suhrkamp.

Part II

MAKING TIME FOR . . . SYNCHRONIZATION

Chapter Five

Making Time, Configuring Life

Smartphone Synchronization and Temporal Orchestration

Martin Hand

Life is busy and steals you away from your personal habits.[1]

The claims made by the "habit tracker" smartphone app Habitify indicate the continued cultural dominance of the notion that we are "too busy." The app promises to "remove distraction and build a stronger routine" so you "know where you are going" in life. To get back on track, users need to classify and subject all their habits (reading a book, learning a language, doing yoga, etc.) to temporal analytics to fully understand where they are going wrong. Habits and routines are thus seen as ontologically distinct from all this distracting busyness, and smartphone apps appear to be the solution. The technological promise here is not to be "free" from habitual drudgery but to regain temporal control *through* habit and routine via devices and data. It is not a question of "saving" time, but remaking time and reconfiguring life. Beyond this commercial hyperbole and personal data extraction, apps such as these suggest that relationships between temporal ordering, digital devices, and the demands of coordinating everyday life have become tangible, visible, and often fraught aspects of everyday experience.

Recent scholarship has illuminated the situated and differential nature of temporalities within networked and mediatized life (Couldry and Hepp, 2017; Gregg, 2018; Keightley, 2013; Sharma, 2014; Wajcman, 2015; Wajcman and Dodd, 2017). The new porosities of work and home (Gregg, 2011), the shifting rhythms and routines of social practices (Southerton, 2009) and the structured temporalities of disadvantage experienced in everyday life (Sharma, 2014) construct variegated images of everyday adjustment, coping, and management of temporal demands. A common theme is that temporal demands produce new anxieties, pushing people to take individual responsibility for coordinating increasing numbers of "desynchronized" activities often using devices and apps now embedded in individual lives through habit (Chun, 2016).

The paradoxes of this are well-known (Schwartz-Cowan, 1983; Rosa, 2013; Shove, 2003; Wacjman, 2015). As Shove (2009, 28) suggests, a "proliferation of autonomy and convenience devices is likely to increase the sum total of arranging required to bring several people, each locked into idiosyncratic schedules, together in the same place at the same time." In this context, the chapter explores the proliferation of efforts at "social synchronization" mediated by personal and distributed calendars, along with scheduling, productivity, and tracking apps that are routinely accessed through digital devices. These are, of course, part of wider developments in evolving socio-technical systems, organizational cultures, work and communication practices, and so forth. But I want to focus here on the microlevel detail of these mediated temporalities, principally discussing individual framings and experiences of temporal management. The research on which this chapter draws asks: To what extent and in what ways do configurations of smartphones and scheduling applications intervene in and restructure the temporality of practices and people's experiences of time?

In discussing this question, the chapter references research drawing on more than eighty in-depth semistructured interviews with professional urban and suburban householders, individuals transitioning to retirement, and university students to examine how temporal expectations are encountered, experienced, and negotiated in the context of sociodemographic differences. These differences included age, proximity to the workplace, gender, and socioeconomic resources characteristic of specific neighborhoods. Interviewees were drawn from a small, a medium, and a large city in Ontario, Canada, purposively recruited through workplace and public advertisement, community groups, and social media. The qualitative analysis identifies four emerging modes of smartphone mediated "temporal orchestration"—accommodating precariousness, coordinating the lives of others, syncing bodies and data on the move, and resisting syncing—that, it is argued, *make time* and *configure life* differently.

SYNCHRONIZATION: TIME, TECHNOLOGY, AND PRACTICE

The "problem" of coordination or synchronization is central to the literature on contemporary temporalities. First, it is an aspect of large-scale changes in temporal organization towards increased fragmentation. Processes and experiences of "speed," "acceleration," and "routinization" in technological modernity (Hassan, 2003; Heidegger, 1999; Nowotny, 1994; Rosa, 2013; Vattimo, 1985; Virilio, 2006) are thought *intensified* in the digital age. This is due to the "economic logic of capitalism" (Rosa, 2005, 448) where time is both compressed and de-sequenced (Castells, 2010). These processes imply substantial changes in how time is structured at the societal, or at least collective, level. In contemporary theories of temporality and media, there are several approaches to synchronization at this scale. For Scannell (2014), "live" broadcasting is a mode of synchronization that produces a "common public time." In terms of digitization and social media, it has been argued that these enable a synchronization of digital labor with the informational circuits of capitalism (Agger, 2011; Fuchs, 2014; Gardiner, 2014). At the most abstract level this turn towards synchronization (of capital, media, persons) articulates how "the current prosthetization of consciousness, the systematic industrialization of the entirety of retentional devices, is an obstacle to the very individuation process of which consciousness consists" (Stiegler, 1998, 4). We might think of the above as concerning "being synced" through media systems, in terms of perception and representation (of "simultaneity") or the "real time" tracking of users.

Second, if we shift to the level of everyday experience, what were previously fixed boundaries between temporal domains are thought to be blurred. Experience is fragmented into ever "smaller" units and the distinction between domains of practice—particularly "work" and "home"—have less clarity and distinctiveness (Gregg, 2011). Sociologically, the relative degree of institutionally and personally timed events is thought to have changed markedly during the postwar period. As Southerton shows, "[th]e temporal rhythms of the contemporary period are characterized by the growing necessity for personal coordination of practices. Collective rhythms and routines of daily life remain, only they are not 'institutionally ordered' in the same way as they were" (2009, 62). A central problem now encountered by many is *how* to coordinate "desynchronized" activities. These changes are not simply driven by the media but have arguably become increasingly complex through the media. On the one hand, there is an identifiable "fragmentation and individualization of the experience of time" (Green, 2002, 254), but also heightened opportunities and expectations of successfully coordinating, scheduling, and synchronizing multiple activities (Southerton, 2003, 2013). Moreover, as the contexts in which digital devices are used and multiplying (home, travel, work, leisure), there are potential tensions between the expectations mediated through technology and the temporal habits and routines associated with those domains (Shove et al., 2012), which may further produce new temporal expectations around the "telemediation of everyday experience" in which we are expected to be "constantly available to and for communication" (Tomlinson, 2007, 158). As Couldry and Hepp usefully observe, "[w]e inhabit a social world characterized by the pluralization of temporalities, on the one hand, and the complexification of technological systems for the coordination of temporality, on the other" (2017, 107).

The theoretical orientation of this research has been a practice-orientated approach to examining how configurations of devices and applications mediate temporality, and how mediated temporal expectations are experienced, negotiated, and managed differently as aspects of ordinary practices. This entails drawing upon approaches that think beyond the specific affordances of digital technologies and consider the ways in which objects and technologies of various kinds are implicated in the temporalities of practice. The temporalities associated with or embedded within technologies must be *realized* within the concrete contexts of everyday life to become significant (Hand, 2016). As Keightley advises, "[t]he analysis of mediated time requires a renewed focus on the way in which time is realized through the reception and uses of media in everyday life and a recognition of the plurality of temporal experiences that this involves" (2013, 71; see also Glennie and Thrift, 2009). Rather than assuming temporal standardization (e.g., "speed"), then, it is equally likely that we will encounter increasing multiplicities and variabilities coupled with efforts to make sense of them. While it would be fair to say that technical processes are faster (e.g., "time critical") and that everyday life appears "busy" to most people, it is quite another thing to assume a direct relationship between the temporal capacities of technologies, a dominant understanding of time, and the actual experience of lived time (see Sharma, 2014).

What does it mean to "make time"? In the first instance this is one of the key vocabularies of temporal experience; one that stresses the active and ongoing practices of temporal production. We "make time" for family, for ourselves, for "getting things done," for relaxation and so forth. What do we mean by this? We might be seeking to step outside of a dominant temporality or universal tempo associated with *being* synchronized via obligations, but in so doing we are engaging in different (temporal) practices. In that way, alternative practice arrangements *generate* temporal experience and make time. As Holmes (2015) argues, it is possible that the nature of the temporal experience becomes the most significant aspect of the practice, rather than only a key feature that shapes and is shaped by it (i.e., enjoying the "slowness" of reading a book or eating; or the "pace" of organizing the work of others). Making time, then, involves the active production of times through an engagement in and orchestration of practices, such that "temporalities are themselves continually reproduced, enacted, and transformed through the sequencing and timing of daily practice" (Shove et al., 2009, 4).

Crucially, this work of sequencing and timing is always-already technical. Indeed, calendars and similar instruments are logistical media that "mediate timekeeping" (Wajcman, 2018a). Technologies also "make time" in this sense, but only in relation to the practices of which they are part and the broader arrangement of devices in which they are embedded. In this chapter, the entry point for discussing temporalities with interviewees is smartphone use. The ordinary use of the smartphone is, I think, the primary way in which individuals directly experience and articulate "deep mediatization" through which "the texture of everyday life [including temporality] is now largely woven out of media materials, the platforms of externalization and exchange" (Couldry and Hepp, 2017, 153). But we have to think of these "platforms" relationally; as contingent elements within multiple practice arrangements (see Crang, 2007). Smartphones are complex, compound, contingent temporalizing devices incorporating several logistical media (clocks, calendars, cameras) and have become recursive elements within almost every other social practice (e.g., eating, exercise, photography, reading, shopping), potentially altering the temporal qualities of those practices, their coordination and the temporal data produced as a result (Hand and Scarlett, 2019). As such, smartphones and their applications enable temporal rescheduling in ways that are part of somewhat diverse, ongoing, negotiable, and interactional efforts at social synchronization.

MUNDANE PRACTICES OF SYNCING

There are many ways in which people understand and adjust to the expectations, demands, and conventions of temporal coordination in everyday life. As noted by several scholars, there are relatively dominant expectations about temporal ordering in contemporary society, that appear to push individuals towards "entrepreneurial" models of time management (Sharma, 2014; Wacjman 2018). Beyond the general sensibility, this means rather different things in diverse contexts of work and institutional cultures, as well being open to diverse forms of interpretation, appropriation, and reorientation in relation to established and continuous temporalities of practice. In what follows, I want to explain some of the mundanities of syncing—efforts to coordinate, schedule, or harmonize elements in time and space—as described by some of the individuals in the research, organized around four modes of temporal orchestration. These are used here to illuminate some differential orientations to synchronization enabled and constrained by situated practices, obligations and expectations around temporality, and the uses of devices and apps in classifying and organizing everyday life.

Accommodating Precariousness

Beyond the world of always connected, urban knowledge workers, the most mundane changes in administrative systems can require individuals to adjust their broader temporal routines. In some cases, this involves changes in the duration of time spent in the office. For dental hygienist Carol, an institutional change in administrative software means that to maintain the essential aspects of her work, she needs to start earlier. This, in turn, means that her "whole day starts earlier," and that a range of previously coordinated practices require minor adjustment and recalibration. Similarly, Mike (a high school teacher) describes how changes in the school board requirements for assessment, and the materialization of that through new student report card software, has been quite transformative in the temporal intensification of those tasks. Commonplace institutional-level administrative changes like these are generally assimilated or "coped with" through the individual adjustment of routine and expectation. For those in relatively stable employment, and without the burden of complex obligations outside the workplace, an admittedly tedious process of adjustment occurs but doesn't necessarily require any radical changes in broader temporal organization or an increased reliance on scheduling systems or apps.

In contrast, those working in multiple "casual" forms of employment, or uncertain part-time service work, increasingly deploy such devices and apps to manage the timing, sequencing, and syncing of practices. This also involves the individual calibration to institutional conditions (i.e., available shifts; numbers of hours) but in ways that must be artfully orchestrated to produce, oxymoronically, a temporary routine. In precarious work of this kind, other aspects of ordinary life—such as finances, health, relationships—are deeply affected by the inequities and uncertainties associated with irregular employment.

An exemplification of how intense such continuous recalibration can be is apparent with Jennifer, a twenty-eight-year-old registered practical nurse. She is a part-time worker in terms of employment status, but stitches together two "casual" jobs with two more regular positions to produce full-time hours over the course of a week. Difficult enough, but it also entails working at two different institutions, and successfully scheduling shift patterns that have twelve temporal variations. To achieve this, Jennifer uses the Shiftworker app on her smartphone. As Jennifer describes:

> So I can put in all the different shifts I can work, so I can work about twelve different shifts, so like a twelve-hour shift, a day shift, a night shift, an eight-hour

> day, a four-hour day a four-hour night, and it just kind of organizes everything, so these are all my . . . every colour means a different shift and they're at different hospitals . . . so I organize it by looking at the icon and the colour and that makes me aware that, yes, I am working at this place at this time . . . the little note tags are all different colours as well 'cause I put notes of when my bills are due.

Of course, the work would be precarious regardless of the devices used to orchestrate all the elements involved, but the intimacy of the smartphone and the affordances of the app lead Jennifer to quantify and calibrate all her other activities and obligations to "fit." She has an active social life, plays sports, and has two dogs to walk. These are all scheduled in the app, but to do this work she has accurately timed these activities. She knows "exactly" how long it takes her to walk to her car, drive to the gym, complete the activity, and get back home. This applies to everything that she does so that she can know whether she "has the time" to engage in these activities. She says that, due to how and when the shifts become available, "I usually know what I'm doing three days ahead of time," and in this sense, she has an extremely limited temporal horizon. This also means that "I don't eat at the same time every day, I just can't fit that into my schedule . . . the dogs have a routine, I don't [laughs]." Jennifer lives in a state of continual adjustment, of ongoing recalibration and efforts to sync all aspects of individual life. The work it takes to simply "go to work" is exhausting. As Jennifer says:

> I feel overwhelmed sometimes . . . mostly because of all the things I have to do to keep things going . . . I'm used to working a lot, but I get overwhelmed because I'm constantly thinking about what I have to do at each time, like I can't just sit and do nothing and not be thinking about it.

What role are digital devices and apps playing here? Like Peters (2013) I suggest that they provide a viable architecture for the orchestration of life, in ways that appear "efficient" and rational. In many ways, what Jennifer is doing is in line with the considerable research on how the mobility of such devices transforms work's intimacy and the temporal density of contemporary work (Gregg, 2011; Wajcman, 2015). But here, most of the work being done outside the workplace has no substantive connection to that employment. The transformation of all activities into color-coded durational blocks to be sequenced and scheduled into a three-day viable *pattern* seems to transform the meaning of those activities, or at least how they are perceived as activities or events that can only take place if they can fit the pattern. Clearly there will be activities which do not fit this pattern. But there is a feeling that they *ought* to be amenable to this. Here, scheduling apps afford daily coping within uncertainty, and entail the broader configuring of life as a sequence of shifting time slots. The timing and sequencing of employment becomes intensely calibrated and organized here—which impinges on much of the day, and transforms other activities into visual temporalized data. This is a primary example of how people adapt to and orient themselves towards specific temporal demands, as Sharma (2014) suggests. This is something like "precision scheduling" (Wacjman 2018), although here it is not primarily for efficiency or productivity gains but to simply "make time" work in ways that can approximate a routine.

Coordinating the Lives of Others

We can see how the demands of coordination and synchronization exert considerable pressure on individuals faced with immediate uncertainties around employment, finance, health, and so forth. Among the interviewees there were several other ways in which synchronization appeared as a central component of daily life, and perhaps even a goal. But the interests at play here are quite varied. This is not only a "practical" problem of trying to coordinate people, things, and places such that events can occur (e.g., a family meal), but involves questions of value, self-worth, and in some cases, becomes a moralizing instrument. For example, one of the most common predicaments faced by our middle-aged participants is the coordination of care for younger and older members of extended families. Here, temporal resources such as "availability" appear stretched across families, as each member may be locked into incommensurable schedules that also must be rationalized and justified in the face of perceived obligations. In other words, the syncing of family member's calendars to "share" the care work equitably (for elderly relatives, say) does not only make time in quantitative sequence (turn-taking), but moralizes duration (how much time have you spent with them?). Geoff, a middle-aged administrator, showed us how he and his siblings coordinate nursing home visits to their elderly father, who has degenerative dementia, through distributed digital calendars, and synchronize this with evidence of physical presence through a print "log book." Their father uses the log book to "remember" who has visited, but each sibling also uses this as a mutual, moralizing surveillance mechanism.

The ways in which these forms of synchronization become visible and subject to judgment take a different form at work. This has been theorized as the "ceding of control" over processes of timetabling, whereby the embeddedness of calendars across software packages utilized in organizations enables employer control and surveillance (Wajcman, 2018). Scheduling devices are also sites of contestation, often involving a significant tension between organizational expectation and individual idiosyncrasy. Within studies of "knowledge workers" and those salaried workers for whom temporal "flexibility" has become a de facto condition, it has been argued that, "Schedules become battle zones of alignment between distributed peers in teams, adding a layer of coordination to what had previously been taken-for-granted office presence" (Gregg, 2018, 6). An important consideration here is that the smartphone's embeddedness facilitates these modes of synchronization across several domains both inside and outside of paid work. This feeling of "perennial context switch" (Gregg, 2018, 7) is articulated clearly among those who try to manage extensive systems of orchestration, seeking to align multiple activities across many spheres. In conjunction with workplace fragmentation and complex scheduling, equal attention must be paid to how other practices—those associated with parenting, fitness regimes, eating patterns, and so on—are similarly multiplying and require orchestration at the level of individual responsibility. It is not simply a question of different modes of copresent interaction facilitated through digitization, but also the kinds of *value* attached to these activities and how this relates to their sequencing and relative ordering.

Similarly, it is not only the copresence of individuals that is of concern here. Smartphones are being used to manage broader interconnected systems of coordination that pull together multiple environments (work, home, cars), people (coworkers, children, partners, friends), and practices (administration, eating, exercise, driving) daily. This is where we see how the articulation and coordination of practices *produces* the broader environment in which others must act; how making time in this way configures individual and collective practice. Taking the case of Shannon (age forty-three), a regional sales manager for a large heating and cooling company, we can get some insight into how the symbolic power of perfect synchronization shapes individual approaches to temporal orchestration through smartphones. The complexities of multiplying contexts, actors, environments, and obligations materialized and *made visible* through the device is evident here.

Shannon must schedule the movements of nine sales staff, such that each can maximize their sales opportunities—being able to meet the most potential customers within a given day. Each of these staff has a schedule that must be synced with the schedules of potential customers, often requiring multiple adjustments during the day as these schedules change; an elasticity facilitated by mobile devices. Shannon uses multiple systems to achieve this. Some of these are synchronized through the Dayforce app, which visualizes staff schedules and allows her to make alterations (a "battle zone of alignment"). In conjunction, there are communications between Shannon and staff through email, message apps, the Calendar app, and a Whatsapp group. Shannon explained that there are different degrees of immediacy associated with each—different temporalities that organize the flows of data and communication (e.g., email is slow, msg is fast, and so on) and the expected sequence of responses. Here we see the coevolution of expectation, skill, and convention through entangled modes of synchronization facilitated through personal devices. As Shannon states "I can't do my job without that phone, so technically I'm working 24/7, or could be . . . I'm on my phone 24/7, always watching for something." Shannon often describes her work as completed in a twenty-four period. That is, a successful synchronization of schedules and sales for a given day is the immediate goal.

The ironies of efficiency are not lost on most people, but this does not prevent the continuation of increased volume and intensity associated with "getting on top" of workloads and aiming to be "better organized." As Shannon states regarding the organizational app used in her work, "[S]o I think you get a whole lot more done, but it also makes more work, like it's a vicious cycle." With two children that are involved in an increasing number of extracurricular activities, the potential fragmentation of schedules and density of obligation becomes difficult to manage. Although such activities are often spatiotemporally fixed, institutionally, they must be still successfully synced with the complex work schedule as they mostly require transportation. The logistics of automobilities (Urry, 2003) here are central to temporal experience. Here, in Shannon's life, we see the proliferation of calendars, reminders, and lists as collective instruments of synchronization configuring daily life in general:

> It's crazy, but I don't need to organize it all for time . . . the day to day stuff, it's not scheduled . . . oh but I have a google calendar as well, that's a whole other thing . . . so we have a family calendar where you both just log in, and this is everything from which kid is doing what, to what colour recycling bin is it this week . . . like I would be lost without this . . . and if it's going to interfere with work at all I will just move it over into my work calendar.

She also suggests that if only "she didn't have to sleep," she would be better organized. She increasingly seeks to make her schedule "find" or "activate" her, through reminders and push notifications:

> A reminder . . . you know what, I do it for everything . . . I set reminders to take my daily vitamin, I just feel like I'm constantly on the go with a gazillion things on my mind, I'm constantly running behind, like I'm always playing catch up for . . . I like it when it preactively notifies me opposed to me having to go and look all the time.

As an example of the active construction of temporal order through digitized means, Shannon exemplifies several aspects common to those undertaking demanding administrative, logistical, and service work that are at the same time occupying several temporally dense environments that are acutely gendered in terms of responsibilities and expectations—most obviously, the orchestration of households and dependents across the family. Her account of "downtime" as involving "either gym work or Facebook" also indicates the degree to which organized activities and the smartphone are embedded in different conceptions and classifications of time. Intense forms of recalibration are taking place here, as schedules across multiple locations are "synced," organizing the sequencing of activities and flows of people. There are multiplying systems (e.g., work office, mobile office, family calendars) with different temporalities that require collective orchestration. It is not only that work bleeds into leisure and vice versa, but that successful coordination increasingly involves datafication, in turn tending towards continual adjustment. Life here is being configured as a successful meshwork of synchronized scheduling systems, mixing mono and polychronic temporal practices.

Syncing Mobile Bodies and Data

There is a self-disciplinary aspect to individual efforts at temporal management, but this appears most intense when configured with mobile self-tracking apps. Individuals might be faced with similar problems of desynchronization and a multiplication of practices to be scheduled, but this is subject to further calibration in terms of synchronizing bodies with data (through controlled exercise, eating) and "outputs" with ideals of productivity while "on the move." These forms of personal analytics are premised on the idea that what bodies do over time can be better measured, analyzed, and improved. Self-tracked data are produced in different ways, within different temporalities of practice. Some are automatically and continuously sensed (heart rate, physical activity, calories burned, sleep patterns) where others are actively input at specific times (calorie and water intake). We might ask whether continuous data about activities in-time invites reflexive awareness of increasingly specific temporal categories, sequences, practices, and their coordination, as many individual activities subject to measurement and temporal scrutiny (Hand and Gorea, 2018). Self-tracking involves rethinking temporality as specific metrics to be acted upon and involves quantifying all activities so that there is no "dead time" or "downtime" that *cannot* be measured. This, in turn, encourages a datafication of everydayness, further intensifying the concentration on the present, making the mundane ("how many steps have I taken"?) permanently manifest on the screen. In this way, we might speculate that self-trackers seek to control and utilize the inescapable fragmentation of time. But in so doing, individualized temporality as a succession of nows becomes fully embedded and embodied in daily routines (walking, commuting, exercising).

Among those who are engaged in individualized syncing (i.e., *primarily* producing and coordinating their own temporal orders) and are individually responsible for syncing collective practices (i.e., of other workers, dependents) there is of course an intense focus on the self. For many younger interviewees, scheduling applications are used in conjunction with self-tracking devices and applications that produce a concentration on the present aimed at producing a future (fitter, more organized, productive) self. Many enact the always-on, fluid, chronoscopic time discussed in much of the literature, but always in relation to specific linear projects of self-disciplining and "improvement." The degree to which self-tracking apps are synced with devices, bodies, and data is variable, often depending on prior practice arrangements, individual concerns about productivity and perceptions of what the data means.

If we take the example of Sarah (age twenty-one), a student, we can see how multiple temporalities coalesce as her devices and apps are enfolded into daily practices. Much like the Habitify example at the outset, Sarah uses her Fitbit app to calibrate exercise and eating routines, but also to render visible "what she is doing." She says,

> But I feel like the Fitbit does sort of contribute to that control feeling, in the sense that I can see what I'm doing and I can be like "well it's fine because I was exercising all day" and I'm a very active person.

In terms of perception, the use of a sleep-tracking app has the further effect of reshaping the day. Self-reflection and temporal management do not end with sleep. Rather, sleep is problematized. It has become a practice (rather than something more like a state) now subject to intervention. As Sarah says, "[s]leep is a big one, it made me confront how little sleep I was getting." She also describes how scheduling, as a mode of orchestrating practice arrangements, proliferates across many spheres of everyday life. The personal calendar is used to "plan like mad," organizing whole semesters through deadlines and then tracking back to set deadlines for "drafts" of schoolwork, and prompts to enable student colleagues to peer-edit these drafts. All schoolwork is timed and sequenced in this way through color-coded blocks and reminders. Beyond that, she suggests that the schedule is the primary frame for most activity:

> I think I'm guilty for trying to fit people into my schedule more than trying to adapt to theirs. So, I guess the people that I do see a lot are like I think they fit into my schedule on top of other things.

Here we see bodily discipline coupled with episodic multiplication, where each discrete temporal moment "produced" by the app is an opportunity for "achievement." For some like Sara, friendships are maintained through scheduling. Life here is configured as a series of goal-meeting actions, where continual data about practices becomes a key resource for self-analysis.

Resisting Syncing

It would be misleading to characterize all the interviewees in the research as chronically busy and digitally "always on," however differentiated. While smartphones are universally owned among the people I have spent time with, there is substantial variation in how present-to-hand they are, how "distracting," and how instrumental they may be in the making and shaping of temporal experience. We have seen how the push and pull of synchronization takes different forms, is underpinned by diverse interests and meanings, articulates value and so forth. But what of those for whom this world of continual adjustment and syncing doesn't make sense? Ideals and practices of "opting out," of disconnecting with the kinds of sociomaterial systems discussed above often take the form of so-called digital detox, mindfulness, or similar self-regulatory approaches to immersion in digital media. To some extent these metaphors have replaced those of the "slow" life. But similarly, the logic here is one of embracing a different temporality. Outside of rather self-conscious (and perhaps short-lived) branded "resistance" to contemporary temporal demands, there are other, subtle, alterative temporalities being made, maintained, or revealed (see Baraister, 2017). We see this quite clearly among some of those approaching retirement, or already retired, where many of the temporal injunctions and conventions associated with paid employment have dissipated. Of course, for some, particularly those in the military, highly regimented linear scheduling has remained intact. As suggested earlier, new technologies are as likely to reproduce as disrupt. But for others, temporal orders and horizons have shifted towards longer durations ("whole afternoons," "seasons") in which temporal experience is expected to be and described as having a different pace. But this still needs to be perceived, made, and orchestrated through the configuration of specific practices.

When the established patterns of days, weeks, months are not subject to chronic uncertainty, or complex planning and orchestration, there are other temporalities that take precedence, such as being "open" to events and surprises. This might be due to living alone, where, soon to retire Nancy says, "on your own, it's pretty easy to be organized." As a nurse coordinator, Nancy is responsible for coordinating the schedules of others, but this only happens during "work hours" and in a fixed physical location. Here we see a normative boundary between "work" and "home" maintained through several mechanisms, one of which is the rather ambivalent role the smartphone plays in the conduct of daily life:

> I don't live with it, I'm totally happy to leave it, I don't feel the need to take it with me everywhere . . . although if I forget it one day I do feel the little pinch of angst . . . but I do shut it off at night.

Nancy can explicitly contrast current temporal orders with prior ones related to when she performed shift work where "you really did need a planner, and then when you had young kids, yeah you had to have some way of keeping track." The significance of location with the life-course is evident here, rather than simply age. Time is made differently now, sometimes through media forms that are considered "slower":

> When I'm not at work I like to take my time getting going, I still like to read a [print] newspaper, and just try to do the things that I want to do. . . . I don't like making a lot of plans, I don't mind a few things, but I think I'm just more of a spur of the moment kind of thing, I don't really like things to be that regimented.

This is not an explicit attempt to "digitally detox," but rather involves a subtle reconfiguring of life through a deflating of planning, a careful management of any porosity that may occur between temporalities, and a normative clarity about what could and *should* be left unplanned. This can often be due to changes in life circumstances (separation, children leaving, etc.), or simply the security of privilege, enabling a self-conscious "taking," "wasting," and "filling in" of time. Life here is the emptying out of temporal obligations, for those without complex (grand)childcare responsibilities of socioeconomic precarity. The nonmedia centric account of everyday life here in some ways reveals the stability of regular employment and little anxiety over productivity. In terms of temporal horizons, there are interesting tensions between the fear of empty time (lack of purpose) and the opening-up of time (spontaneity). About the former, Nancy says, "I have a sense of retirement, and I'm not sure how that'll look, I'll have to find something to do, I can't imagine having nothing all of a sudden." To configure life in terms of the latter, she aims to schedule the unscheduled to "create the chance of spontaneity," by inserting deliberate "empty slots" in her calendar.

CONCLUSION

> Logistical media are ordering devices; part of the infrastructure or scaffolding that configures arrangements among people and things. Digital calendars increasingly play such a role, setting the rhythm of everyday timekeeping practices that inform our consciousness of temporality. (Wajcman 2018, 14)

Over the course of this exploratory chapter, I have aimed to show how ordinary efforts to coordinate and synchronize everyday life increasingly involve smartphone orientated modes of temporal orchestration. The four interrelated modes of temporal orchestration described above are not individually produced, but neither are they simply reactive or responsive to dominant temporalities instantiated through devices and instruments. The active negotiation and production of times is both individually and collectively orchestrated through complex configurations of people, things, values, and environments. The roles of digital devices and applications, as logistical media, are of course partly the outcome of how they are designed and facilitate specific orientations to time and temporality. But, as suggested earlier, they must be put to use, and we can begin to see how the metaphor of *orchestration* sheds some light on how individual and collective projects that have existing social-material, practical, and ethical dimensions contextualize the significance of media forms. Digital calendars and the like might have been designed "by and for knowledge workers" (Wajcman 2018), but the ways in which those assumptions about temporality are subject to *ongoing* processes of reconfiguring within quite different material-symbolic contexts raises questions about what characterizes the locally experienced tensions between technology, time, and everyday practice.

Following Sharma (2014), these modes of temporal orchestration might be usefully conceptualized in terms of "recalibrations," stressing the ways in which often contradictory temporal expectations are differentially encountered, and precipitate alternative efforts to synchronize (or embrace desynchronized) elements of daily life. Efforts to actively synchronize involve others being synced, and the ability to remake time and configure life "successfully" through media is unevenly distributed, and may itself produce new temporal expectations, anxieties, and pressures. In this sense, "calendar work" is not always the positive management of time; it is sometimes situated "coping" and "stitching." In terms of temporality and power, it is crucial to recognize the differentially structured sources of temporal expectation encountered by individuals, especially in relation to gendered conceptions of *responsibility* for temporal coordination (e.g., in the case of Shannon). The individualized responsibility—materialized through the personal mobile device—is also an outcome of existing rhythms anchored in gendered practice (Hochschild, 1997). Similarly, the ways in which life-course events and transitions can have a considerable effect on temporal experience, expectations, and moral and ethical commitments to the organization of temporalities and the obligations to synchronize appears similarly significant. This seems crucial in understanding both the variability of what temporalizing devices do, and the shifting sources of temporal expectation, obligation, and value that shape individual's orientation to practices (see Holmes, 2015). For some, syncing primarily involves orchestrating the body in relation to data, where for others it might be the maintenance of an extended network of care relations. While both may involve, at the most general level, a conception of life as "busy" and time as "short," the ways in which time is (re)made through quite different orientations is significant.

From the observations and discussions above it is difficult to disagree with Sharma's (2014) suggestion that all this emphasis on time management leaves people in a state of "constant marginal dissatisfaction." For most people interviewed, temporal coordination is a practical problem, one that produces additional time pressures, and is both bound up with and productive of new expectations for how life should be configured. I suggest that the smartphone is transformative though clearly not in any uniform way. So, what is distinctive about these media here? First, the embeddedness of smartphones in everyday life means that *multiple* logistical media (clocks, calendars, cameras, reminders) tend to become default ways of perceiving, representing, classifying, and ordering time. Smartphones enable the simultaneous codification and *visualization* of multiple temporalities (e.g., sequences, duration, timings, orders) and thus "make time" at several scales. This can mean that people's senses of how "organized" they are, what their temporal horizon is, and whether they "have" time are being shaped by the peculiarities of visual data. Second, the sheer variability of these compound media as they become elements or "relays" in diverse practice arrangements means that there are increased possibilities for organizing time differently, and this can be both liberating and oppressive depending on sociomaterial circumstances. But rather than being simply habituated to mediated temporalities (as if these were uniform), we can see both intensifications of existing temporalities and opportunities for disruption and rethinking emerging.

NOTE

1. Habitify, https://www.habitify.me/, last accessed January 16, 2020.

REFERENCES

Agger, Ben. 2011. "iTime: Labour and Life in a Smartphone Era." *Time and Society*, 20, no. 1: 119–36.

Baraitser, Lisa. 2017. *Enduring Time*. London: Bloomsbury.

Castells, Manuel. 2010. *The Rise of the Network Society*. Oxford: Blackwell.

Chun, Wendy, H. K. 2016. *Habitual Media: Updating to Remain the Same.* Cambridge, MA: MIT Press.
Couldry, Nick. 2012. *Media, Society, World.* Cambridge: Polity.
Couldry, Nick, and Andreas Hepp. 2017. *The Mediated Construction of Reality.* Cambridge: Polity.
Crang, Mike. 2007. "Speed = Distance/Time: Chronotopographies of Action," in *24/7* edited by Robert Hassan and Ronald E. Purser, 62–88, Stanford, CA: Stanford University Press.
Fuchs, Christian. 2014. "Digital Prosumption Labour on Social Media in the Context of the Capitalist Regime of Time." *Time and Society*, 23, no. 1: 97–123.
Gardiner, Michael. 2014. "The Multitude Strikes Back? Boredom in an Age of Semiocapitalism." *New Formations*, 82, no. 2: 31–48.
Glennie, Paul, and Nigel Thrift, N. 2009. *Shaping the Day: A History of Timekeeping in England and Wales 1300–1800.* Oxford: Oxford University Press.
Gregg, Melissa. 2011. *Work's Intimacy.* Cambridge, UK: Polity Press.
———. 2018. *Counterproductive.* Durham, NC: Duke University Press.
Green, Nicola. 2002. "On the Move: Technology, Mobility, and the Mediation of Social Time and Space." *The Information Society*, 18: 281–92.
Hand, Martin. 2016. "Persistent Traces, Potential Memories: Smartphones and the Negotiation of Visual, Locative and Textual Data in Personal Life." *Convergence*, 22, no. 3: 269–86.
Hand, Martin, and Michelle Gorea. 2018. "Digital Traces and Personal Analytics: iTime, Self-Tracking, and the Temporalities of Practice." *International Journal of Communication*, 12: 666–82.
Hand, Martin, and Ashley Scarlett. 2019. "Habitual Photography," in *The Routledge Companion to Photography Theory*, edited by Mark Durden and Jane Tormey, 410–25, London: Routledge.
Hassan, Robert. 2003. *The Chronoscopic Society: Globalization, Time and Knowledge in the Network Economy.* Bern: Peter Lang International Academic Publishers.
Heidegger, Martin. 1995. *The Fundamental Concepts of Metaphysics.* Trans. William McNeill and Nicholas Walker. Bloomington: Indiana University Press.
Hochschild, Arlie. 1997. *The Time Bind: When Work Becomes Home and Home Becomes Work.* New York: Metropolitan Press, 1997.
Holmes, Helen. 2015. "Self-time: The Importance of Temporal Experience within Practice." *Time and Society*, 27, no. 2: 176–94.
Keightley, Emily. 2013. "From Immediacy to Intermediacy: The Mediation of Lived Time." *Time and Society*, 22, no. 1: 55–75.
Nowotny, H. 1994. *Time: The Modern and Postmodern Experience.* Cambridge: Polity.
Peters, John Durham. 2013. "Calendar, Clock, Tower," in *Deus in Machina: Religion and Technology in Historical Perspective*, edited by John Stolow, 25–42. New York: Fordham University Press.
Rosa, Hartmut. 2013. *Social Acceleration: A New Theory of Modernity.* New York: Columbia University Press.

———. 2005. “The Speed of Global Flows and the Pace of Democratic Politics.” *New Political Science*, 27, no. 4: 445–59.

Scannell, Paddy. 2014. *Television and the Meaning of “Live”: An Enquiry into the Human Situation*. Cambridge: Polity.

Schwartz-Cowan, Ruth. 1983. *More Work for Mother*. New York: Basic Books.

Sharma, Sarah. 2014. *In the Meantime*. Durham, NC: Duke University Press.

Shove, Elizabeth. 2003. *Comfort, Cleanliness, and Convenience*. Oxford: Berg.

———. 2009. “Everyday Practice and the Production and Consumption of Time,” in *Time, Consumption and Everyday Life: Practice, Materiality and Culture,* edited by Elizabeth Shove, Frank Trentmann and Richard Wilk, 17–34. Oxford: Berg.

Shove, Elizabeth, Matthew Watson, and Mika Pantzar. 2012. *The Dynamics of Social Practice: Everyday Life and How It Changes*. London: Sage.

Southerton, Dale. 2003. “Squeezing Time: Allocating Practices, Coordinating Networks and Scheduling Society.” *Time and Society,* 12, no.1: 5–25.

———. 2009. “Re-ordering Temporal Rhythms: Coordinating Daily Practices in the UK in 1937 and 2000,” in *Time, Consumption and Everyday Life: Practice, Materiality and Culture,* edited by Elizabeth Shove, Frank Trentmann and Richard Wilk, 49–63. Oxford: Berg.

———. 2013. “Habits, Routines and Temporalities of Consumption: From Individual Behaviours to the Reproduction of Everyday Practices.” *Time and Society*, 22, no. 3: 335–55.

Stiegler, Bernard. 1998. *Technics and Time, 1*. Stanford, CA: Stanford University Press.

Tomlinson, John. 2007. *The Culture of Speed: The Coming of Immediacy*. London: Sage.

Urry, John. 2000. *Sociology beyond Societies: Mobilities for the Twenty-First Century*. London: Routledge.

Vattimo, Gianni. 1985. *The End of Modernity*. Baltimore, MD: John’s Hopkins.

Virilio, Paul. (2006 [1977]) *Speed and Politics*. Cambridge: Semiotext (e).

Wajcman, Judy. 2008. “Life in the Fast Lane? Towards a Sociology of Technology and Time.” *The British Journal of Sociology,* 59, no. 1: 59–74.

———. 2015. *Pressed for Time: The Acceleration of Life in Digital Capitalism*. Chicago: The University of Chicago Press.

———. 2018. “The Digital Architecture of Time Management.” *Science, Technology, & Human Values*, 44, no. 2: 315–37

———. 2019. “How Silicon Valley Sets Time.” *New Media & Society*, 21, no. 6: 1272–89.

Wajcman, Judy, and Nigel Dodd, eds. 2017. *The Sociology of Speed*. Oxford: Oxford University Press.

Chapter Six

Everyday Time Travel

Temporal Mobility and Multitemporality with Smartphones

Roxana Moroșanu Firth,
Sean Rintel, and Abigail Sellen

How have you last checked the time? Chances are that you did not need a specialized technology, such as a watch or a clock, because the device you are using to read this text on might display the time on one corner of the screen or another. These days it is almost impossible to not know the time. Most, if not all, digital technologies, from fitness trackers to smart TVs, have an incorporated time feature that shows the satellite-determined exact time. This chapter argues that while situating everyday life in relation to a globally precise time, digital technologies can also be creatively employed to unsettle time—to jump from present to past and future, or to juxtapose moments to create unique temporal experiences. The accounts discussed here focus on the temporal experiences of people in full-time employment working in the fast-paced sector of technology. In these contexts, the tool that is specifically employed to navigate and manipulate time is the smartphone.

Both the multiplicity of affordances, and the physical characteristics of the smartphone, make this the preferred tool for acting upon time. First, the smartphone affords a multifaceted temporal orientation through storing and giving access to content that allows the user to connect to various points in time, for example in actions of anticipation and reminiscing. Digital calendars provide a visual platform for planning, coordinating, and imagining the near future. Social media and news notifications create the sense of a shared present by instantly connecting people going through their daily activities to global media events. Expressive and functional pictures and screenshots on mobile devices provide new opportunities for action on multiple time-scales, from long-term archival to use in future actions. This situation where a single tool provides access to multiple points in time is unprecedented, and it implies transformation of social relations and notions of time. Second, the physical characteristics of the smartphone, such as the fact that it can easily be carried around at all times, makes it into a tool to use in any situation. The smartphone combines both aspects of a tool that is "ready-to-hand" and "present-at-hand" in Heidegger's phenomenology (Dourish, 2001). Ready-to-hand means that we act through the tool not paying attention to the fact that it mediates our action, for example when we just pull it out to check an email. Present-at-hand means that we act upon the tool to achieve the desired effect, for example when we lift our arm up, holding the phone in search for a suitable wireless signal. By combining both aspects in our use of the smartphone we become aware of its capacities and limitations and we are able to employ it creatively in a variety of situations; for example taking a photo of a page in a book to create an extra copy of a piece of information that is needed in the moment, or anticipated to be needed in the future.

New technologies that support simultaneity in the circulation of data flows have been regarded, in the last two decades, as generating a new temporality of acceleration, also described as "timeless time," "time-space compression," or "instantaneous time" (Castells, 1996; Eriksen, 2001; Harvey, 1990; Rosa, 2013; Urry, 2000; Virilio, 2012). The novel pressures brought by constant connectivity have led to demands for greater productivity and to the normalization of work extension into home and leisure time (Gregg, 2018; Mullan and Wajcman, 2019). These changes are summarized by two main claims: that we live in times defined by speed, with regards to sociotechnical transformations, and also to the pace of everyday life; and that the best way to deal with time scarcity is intelligent scheduling (Nowotny, 2005; Rosa, 2013; Wajcman, 2019; Wajcman and Dodd, 2017), at least for the ones who can afford it (Sharma, 2014). Both claims describe very well the work and lifestyle patterns that the people on whose accounts we report here are enrolled in. However, these circumstances are just a background to their development of creative tactics for stepping aside a time that is continually speeding, and for creating, instead, a wide range of experiences of time.

This chapter contributes to a growing literature addressing the diversity of contemporary experiences of time (Adam, 2004; Bear, 2014; Glennie and Thrift, 2009; Greenhouse, 1996; Guyer, 2007; Keightley, 2013; Wajcman, 2015). While new technologies might contribute to reshaping people's sense of time, they can also be employed as tools to act upon time—to rewind or slow it down, to disassemble and reassemble it in new forms and interpretations. According to Birth (2012), all cultural systems of time function by mixing ideas of time, as well as the tools for measuring it. The capacity of digital technologies to automatically connect to satellites to determine the exact time does not subsume the problem we inherited in the construction of calendars regarding the difference between the lunar year and the solar year. In fact, we have always used a multiplicity of tools for measuring, and for making sense of time. While individual experiences of time might be conditioned by the narrative of acceleration, this does not mean that they are always overruled by it. Here we unpack such individual experiences, shedding light upon the creativity and agency of people who are acting upon the temporal conditions they encounter by developing tactics of "time-tricking" (Moroșanu and Ringel, 2016).

In the accounts discussed here, the preferred tool for navigating and manipulating time is the smartphone. We approach the smartphone analytically as a cognitive artefact and time-reckoning tool (Birth, 2012). This means that it is an artefact in which humans have placed knowledge, especially, for our purpose here, knowledge about telling the time. The smartphone, like other digital technologies, can tell and display the time with exact precision. As people come to rely on cognitive artefacts, they do not need to continue to cultivate the skills for producing that knowledge themselves. We do not need to be able to tell the time by observing the sky because we can find the time in our pocket. However, people are very good at developing vernacular approaches to technologies—at using them in creative ways for lots of purposes that were not foreseen when the device was initially developed. Digital anthropologists have described such vernacular practices with regards to a wide range of technologies, from radio (Tacchi, 2000), to cell phones (Horst and Miller, 2006), to mixes of communication technologies (Madianou and Miller, 2013). While the smartphone is a cognitive artefact embedding a dominant model of temporal linearity, people are still able to use it creatively in reinterpreting and remixing points in time, tenses, speeds, and durations.

Our discussion focuses on two types of orientations when acting upon time: temporal mobility and media multitemporality. Temporal mobility means moving between distinct points in time, such as jumping from the present to the past. This is when one wishes to escape the present moment, for example in situations that involve commuting, traveling, or waiting. With a smartphone it is possible to jump back to the past, by immersing into browsing one's photo gallery that might go back up to several years.

In the second orientation, that we call media multitemporality, the past, present, and future are not navigated as distinct points in time, but they become juxtaposed. The philosopher Michel Serres introduced the term "multitemporality" to refer to an understanding of time as containing multiple pleats. This corresponded to his method of inquiry that consisted of placing works from different eras and different disciplines in dialogue (Serres, 1982; Serres and Latour, 1995). In the accounts discussed here, the smartphone allows the juxtaposition of the past, present, and future. They follow an understanding that past moments accumulate a certain emotional value in time that cannot be grasped or foreseen when those moments are originally experienced—when they are present. With the view of being able in the future to access the value that at present is illusive, the participants reported recording ordinary details of their lives, as well as work outputs, that they would otherwise overlook.

This chapter brings together ethnographic findings from two separate studies. The first study looked at digital media practices in relation to energy demand in UK homes between 2011 and 2014, and it was part of a wider interdisciplinary research project addressing digital interventions for reducing energy demand. The second study investigated functional mobile image capture in the workplace, and it was conducted in 2018 in Cambridge, UK, as part of an industry placement research fellowship. Here we focus on findings regarding temporal navigation via smartphones, which are common to both studies. While the first study took place in domestic settings, the second study was set in the workplace. By bringing these findings together we therefore address the ways in which smartphones are employed to act upon time across settings. We describe the studies in more detail below, in the two sections that are dedicated to temporal mobility, and, respectively, media multitemporality. Throughout this chapter we use pseudonyms when referring to individual research participants.

"NORMALLY, TIME IS SOMETHING THAT WE DON'T HAVE ENOUGH OF ANYWAY": TEMPORAL MOBILITY AND TIME SCARCITY

The ethnographic study looking at digital media practices and energy demand was set in homes in a small town situated in the Midlands area of the United Kingdom. It involved long-term fieldwork with twenty families using a mix of methods from more traditional ethnographic methods, such as semistructured interviews and participant observation, to visual and arts-based methods that were specifically designed to address the topic of time and digital media (Moroşanu, 2016a).

For example, the Tactile Time Collage asked the participants to create an artwork that represented their media practices around the family TV. The collage below, created by Cynthia and her family, places the TV at the centre of a clock face, drawing out all the time segments when the device is on (figure 6.1). It includes other devices that are employed concomitantly, such as smartphones and laptops; and food and beverage items that accompany the usage, such as toast and cups of tea. The routine of each family member is represented by their initials. The collage also includes small pieces of fabric that provide a tactile representation of what each segment of time in the collage feels like for each family member. For example, the daughter chose a

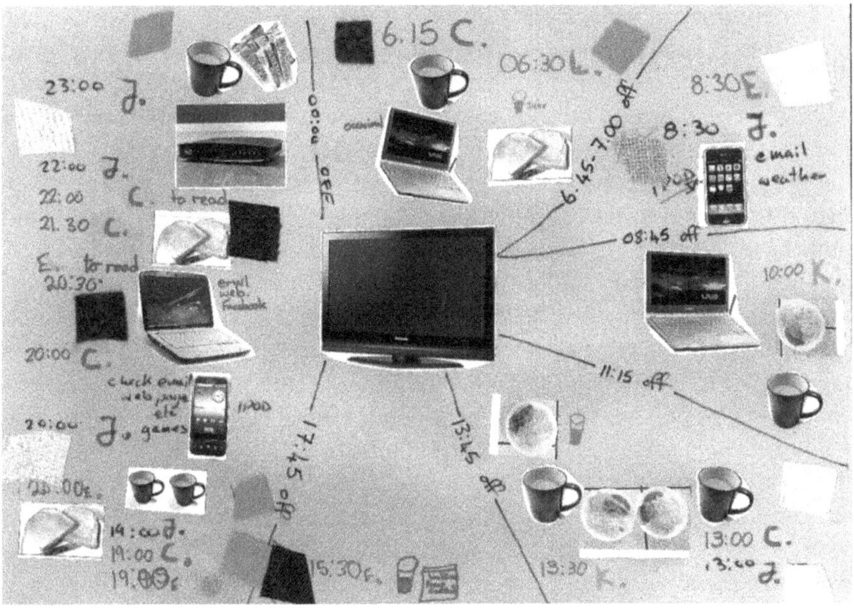

Figure 6.1. Tactile Time Collage made by Cynthia and her family

soft fleece fabric to show that this is how she feels in the early morning while having her breakfast in front of the TV; while the dad chose hopsack to illustrate the segment of time when he is checking his email and the weather forecast just before leaving for work.

Such multidevice intersections, for example when people checked work-related issues while sharing a physical space and media background with their family, were common in this study. It has been estimated that a total volume of eleven hours of media and communication activities undertaken by an adult per day in the United Kingdom is squeezed into eight hours and forty-one minutes because people are often engaged in two or more media activities at the same time, also referred to as media multitasking (Ofcom, 2014). The portability of smartphones means that people can extend their multitasking practices to other situations and devices. For example, Chris said that he cannot bear having to wait thirty seconds to warm up his cup of tea in the microwave, so during that time he checks his calendar for the coming weeks on his phone, or he looks at his email. The smartphone allows him to escape a situation of boredom by looking at what the future holds regarding scheduled activities, or by doing a time trade that involves quickly reading some emails in order to then shorten the time he spends on his computer. In these cases, the smartphone affords what we call temporal mobility—the flexibility to move between distinct points in time.

Temporal mobility was mentioned in the Cambridge study as well. The participants reported that when commuting, traveling, or waiting they often used their phones to remember happy times, by looking at photos from holidays, or to explore an interest that they hoped to materialize in the future. For example, Isabelle described how she usually spends her time while on the train, searching and selecting images for products she is interested in: "So, this is an industrial light I'm entertaining to put into my living room. Sometimes I sit on the train and browse and think 'Okay, this one I like.' So I take a screenshot and keep it aside." The screenshot is part of a list of images of products that Isabelle curates with the view of consulting later, and eventually making a decision. Having a smartphone that streams data regardless of Isabelle's physical location means that she can spontaneously engage in curating her list while on the train. This keeps Isabelle entertained while traveling, but it also affirms an interest that she wishes to explore in the future. She projects her attention to a future that she is excited about, leaving behind a dull present moment. Similarly, she recalls a conversation with her boyfriend via a messaging app where they debated ideas for weekend trips. Reading about a photography exhibition in London, they were prompted to search and then share each other's favorite photograph from the online exhibition brochure. As a preparation to visiting the exhibition as a couple, they had already tried out a way of finding more about each other by sharing miniature digital versions of the photos they anticipated viewing holding hands.

Actions of temporal mobility facilitated by smartphones reflect the participants' wishes to escape a present moment that is not particularly enjoyable, but they also suggest time scarcity. There is a sense that by doing a specific task now they might be able to free some time for later, such as in the example of Chris doing a time trade. Time scarcity means that people are not able to do all the activities they are interested in. At the time of the interview, Isabelle had not visited the photography exhibition yet. While being aware of time scarcity, the participants in both studies expressed the wish to keep cultivating varied interests rather than conceding to uneventful and predictable routines. Comparing his family's work and lifestyle patterns with his parents' generation, Chris says that his evenings are not aligned to repetitive routines, but they bring a bundle of varied tasks that need to be completed, from supervising children's homework to arranging next day's work trip:

> Normally, time is something that we don't have enough of anyway [. . .] I guess the modern world and family processes, they don't really lend themselves to the kind of relaxed, context repetition of previous generations. I mean, we're just doing whatever we need to do to get everything done at the moment, aren't we?

Time scarcity and the feeling of always being rushed provided a background that encouraged actions of temporal mobility. Evenings often involved jumping from being present in the moment and enjoying family time, to planning work tasks for the next day. One participant in the first study admitted that he kept checking his email through the night, between sleep cycles, in order to send swift replies to work partners located in other parts of the world with significant time differences, such as Asia. However, most participants referred to a temporal threshold that marked the definitive end of their working day, when they would definitely stop checking their work email. This tended to be 10 o'clock in the evening, coinciding with the late evening news bulletin. Nicholas, a participant in the second study, reflects on his habit of keeping a set time that symbolizes the end of the working day:

> So I have, kind of, a thing, as it were, when the clock changes to 22:22 at night, quite often I'll take a screenshot of that. It's quite significant. I'll have hundreds of them as a general rule, and I'll send them to various friends. It's almost a marker that the day is done, "Well done, you've nailed it."

Having made it through another day of meetings, research, and presentations at his senior job for a tech company, Nicholas feels the need to celebrate by sharing with his friends, via a messaging app, the realization that the day has ended. The moment that he shares is not related to what he is doing at the time, as one of his friends, for example, sends photos of her feet up on a footstool, relaxing. What he shares is the moment when the clock on his phone changes to a numeric pattern that for him signifies the end of the day, while also displaying a distinctive visual feature. When using digital clocks that follow a twenty-four-hour display, the visual representation of 22:22 is screened only once a day, for sixty seconds. It is a relatively rare moment when numbers are aligned in such a distinctive pattern, and, more significantly, it is a moment that Nicholas can observe.

The counterpart to this time is 11:11, which is part of the work hours and, presumably, during the busiest time of the day. Clock changes in the day signify activities—the next scheduled meeting, the departure time of a train. In the evening, however, it is possible to notice the clock change without having to think about the next activity. Time is no longer a constraining factor that one needs to fight against, but it becomes externalized—a flow whose changes are noticeable, just as it is possible to observe the wind or the rain. The fact that in recent years the visual representation of time has shifted from analogue clock faces to digital numeric representations that are featured on almost any digital device, from fitness trackers to smart fridges, is not unrelated to a growing sense of time scarcity, as argued elsewhere (Moroşanu,

2016a). While analogue clock faces allow the viewer to anticipate the next hour and literally watch the time go by in repetitive cycles, digital clocks only show the exact current time; similarly to a stopwatch, the visual counting up in a linear progression displayed on digital clocks encourages an imagination of time as something to be racing against. In the example given by Nicholas, 22:22 marks the finish line. That is the end of his working day, but also the moment when time measurement ceases—"a marker that the day is done." What follows, until the next morning, is a segment of life that one is not required to juxtapose over a numerical timeframe.

"THINGS THAT YOU THINK ARE BORING EVERY DAY AREN'T IN THE FUTURE": MULTITEMPORALITY AND FUTURE NOSTALGIA

The second study we report upon in this chapter looked at image capture for information—the ways in which people use smartphones to take pictures or screenshots of informative content. It was conducted in Cambridge, United Kingdom, and it included semistructured interviews with nineteen participants and, in some cases, workplace tours addressing some of the situations, and locations, where the participants captured functional photos. From the overall sample, a subsample of twelve participants worked for technology companies, which represent a large part of the industry sector in Cambridge. Here we focus on the findings that emerged from working with this subsample.

We found out that in the functional photos people took and kept, there was often a connection with temporal frameworks and experiences. These images included smartphone screenshots of snippets of information to be used or shared with friends in the near future, such as train times or cinema listings; pictures that the users took of their shortlisted preferences to be perused and decided upon later, such as labels displaying the price and the dimensions of rugs viewed in a shop; photos that assisted with completing a task in the moment, such as the reading of an unreachable gas meter located high up in a cupboard that the user had to write down and send to their gas provider; pictures of achievements to be included in a long-term collection, such as cakes one has baked, or completed puzzles one has solved; snapshots of serendipitous encounters, with neighborhood cats, or funny signs on city streets, that one wanted to remember. This variety of information was recorded and stored on smartphones for virtually forever, as the memory capacity of the device, or the costs of cloud storage, were not considered to be limiting factors. The number of pictures that people stored on their phones was more than twenty thousand in some cases. These archives were often consulted in search for specific information, and sometimes casually browsed for passing the time.

One category of photos was present in most of the accounts, and it highlighted the participants' different expectations regarding the future. This consisted of pictures of information written down on whiteboards. At the end of meetings that included scribbling down solutions or ideas on a whiteboard, the participants took pictures of the resulting text and sketches. This was sometimes for a clear purpose. For example, Steven, a computer scientist, recalled that he would capture the information on whiteboards in a meeting room when realizing that his team's time was up, and the group that booked the room for the next time-slot was waiting at the door. He would use that picture as a resource when, back to his desk, he drew out the action points.

In other cases, the participants took pictures of whiteboards with no exact purpose, but with a general expectation that the information might prove valuable in the future. Sarah, a software engineer, reported that she would take whiteboard pictures during workshops if other people in the room did. She explained: "I'd think they'd seen something of value that they wanted to keep for later, so I might as well. You don't know what you'll need in the future." The fact that the field that she is working in is moving fast, means that any information and ideas emerging from workshop interactions might become useful at some point, in the near or far future. Sarah would not take the risk of missing out on information that her colleagues are recording.

Dave, a researcher in human-computer interaction, recounted similar practices of collecting with the view of future needs. He described himself as "a bit of a squirrel" because he keeps all the photos he has ever taken. His reasons for taking whiteboard pictures were somehow different. While he knew from experience that he might never use those pictures again, he still wanted to store them for sentimental reasons:

> There's a lot of stuff where it's relevant in the moment. I take a photograph of it that I'm probably never going to use for anything. But it's still nice to have a record of it, because when you've worked on something for a while it feels wrong to just throw it away immediately.

By creating a record of the workshop outputs, Dave tries to extend the present moment and affirm the efforts of the workshop participants as meaningful. At the end of the workshop, following the excitement of discussing and bouncing off ideas, the outputs feel relevant. This moment is extended through the act of taking a picture. Whether or not the workshop results would be appreciated or taken forward by the management, for Dave there is still value in the time he spent working on that topic.

In these examples of people taking pictures of whiteboards, the present moment is evaluated in relation to imaginations of the future. If recorded correctly, the present might provide an answer to future questions and needs that one cannot yet foresee. These are cases of what we call media multitemporality, following the concept of multitemporality developed by Serres (1982). The past, present, and future are here juxtaposed in the action of taking a picture. An image of workshop outputs represents a record for the future, it pays tribute to the work undertaken in the very recent past, and it brings additional meaning to the present, as the moment when a potential solution to a future question has been created. This juxtaposition is achieved with a single tool—the smartphone—which can be employed for taking pictures, distributing them to everyone in the room, and storing them. Had the tool been missing, or not displayed these affordances, Sarah and Dave might not have cared about documenting the workshop outputs. One can argue that the access to, and ease of use of the tool, supports approaches to time that involve juxtapositions of past, present, and future—or media multitemporality.

By using a smartphone, it is possible to create an individual archive that includes the thoughts and ideas one has been exposed to over the years and the situations where one has imagined solutions and scenarios. These potential ideas for the future are then collected alongside photos of food one has made, and cats one has met, in a surreal collage of moments one has lived. Here the smartphone emerges as a tool for curating and maintaining an archive of the self. A designer working for a tech company, Umesh had lived in six different cities by the time he was thirty-one. Below he talks about the ways in which he consults the photo archive on his phone:

> Umesh: I have pictures from 2011 or 2012 here, and sometimes, when I don't have anything better to do, I just look back and I remember that day, or whatever. I lived in six different cities for a considerable amount of time in each, so sometimes I just look for the city name and it shows me where I was. I was in Bangalore for almost four years, so I just search for Bangalore and it shows me all photos from there. And that's a particular time that I spent—starting from friends, to some forms that I filled in, to screenshots. But I like having all these. Like when I first got my iMac, to some medicine that I bought and why I bought that [laughing].
>
> Roxana: So, do you remember the moment when you took this photo?
>
> Umesh: I don't remember it without seeing it. This acts as a trigger for me. This is something I was working on [photo of computer screen], then photos of my workplace. Because I don't work there anymore, but I want to remember, so that's why I have pictures.

The pictures Umesh enjoys browsing represent everyday details that when brought together recreate a certain period in his life. They are not photos of events, but evidence of his day-to-day preoccupations and experiences. The moments, and things, that were significant enough, for various reasons, for him to capture at the time, help him to remember the ordinariness of life in a different part of the world. This practice of reminiscing by browsing his camera roll counteracts the overarching dynamic of constant moving from one place to another following career opportunities. Scattered over six cities around the globe, his memories and experiences can be collected in one place—his phone—that does not leave room for error. If his memory might ever play tricks, for example confusing the Bangalore with the Bangkok workplace, he would be able to tell them apart by checking the pictures stored on his phone. When he feels like reminiscing, Umesh can consult this archive of the self, taking his time to reflect on past moments that otherwise might get lost in the accelerated work and lifestyle patterns he is enrolled in.

Sarah describes a similar orientation to documenting the familiar and ordinary details of life:

> I've started doing One Second Every Day, which is a video app, where you take one second clips of video, and then you can go back and see it, which I quite like for the nostalgia aspect. But, what I actually realised about it, is that things that you think are boring every day, aren't in the future. So, if I go back and look, I've been doing it for over a year now. A year ago I was living in a different house with a different group of people, and there's just all these little things, which at the time I was like, "Oh, I should just take a picture of something," and it was boring. And now I'm like, "Oh, yeah, the kitchen looked like that," and "We used to watch that TV show," and stuff like that. So, I try and keep stuff that's just random pictures because in the future it might mean more than it does right now. For example, if I take a picture of my office right now, I'd be like, "Why would I want to look at that? I'm in my office for eight hours every day." But if I moved jobs, even if I moved to a different office in the building, I might be like, "Oh yeah, that's what I had in my office." It's a nice feeling, so I'm kind of prepping myself for that feeling in the future.

What Sarah describes here is how she has learnt to appreciate visual representations of what she previously considered to be routine and noneventful occurrences, "boring things" as she puts it. The value of these visual records is to be recognized in the future, after one has moved or changed routines, and therefore altered the texture of their everyday life. When taking "random pictures," for example of her office, Sarah anticipates her future nostalgia. She prepares for the nice feeling that comes with nostalgia by creating the conditions and the visual material that would later trigger it. When taking the picture, she conveys additional meaning to an everyday setting by imagining how she would look at the visual representation of this setting in the future. In this example, by being juxtaposed with the future in an action of media multitemporality, the present achieves a higher status. The present is not just an ordinary moment, but it also is the future past. When one is in a rush and cannot stop to fully appreciate the present, one can always take a photo to return to and reflect upon later. The smartphone allows for rewinding as well as forwarding, and any other customised mixes of points in time that one might wish to create.

CONCLUSION

The approaches to time described in this chapter are certainly not timeless. They are situated within a framework of speeding and time scarcity that requires clever solutions for one to succeed at work while still maintaining a sense of enjoyment of one's life. The people we talked to are striving to make it, to beat time, and constantly prove that they are up to date with the contemporary sociotechnical changes affecting their area of work. Like the tech workers described by Wajcman (2019), who are watching and listening to content at double speed so that they can cover more information and be savvy about new trends, our research participants have their own creative techniques for riding the speeding waves of time. They are not trying to evade time, but are finding ways of working with, and within, acceleration.

In the accounts we discussed, the smartphone is a multifaceted tool that can serve almost any purpose. People use it for work, for keeping in touch with their loved ones, and for entertainment. Similarly, it can connect to almost any point in time—future, past, and present. It is employed for scheduling future meetings, browsing through old photos that bring back happy memories, and sharing information in the moment, such as live location. Because of its multifaceted nature that covers all these areas of life, the smartphone is a tool that can be used to curate and maintain an archive of the self that is almost all-encompassing. The information stored on a phone, from photos to apps, does not in the least represent solely a work archive, just as an individual cannot be reduced to just one of their social roles and interests.

However, if during times of acceleration, a single aspect of life, such as work, might sometimes take primacy, people are able to remind themselves that they are complex individuals, and not just workers, by browsing through the evidence in their phones. That is, evidence of their lives that are not lived for single and clear purposes alone, but include happenstances that make no sense, moments of pain, or joy, that appear unexpectedly, serendipitous encounters that might provoke laughter, or change one's life course. People access and enact this complexity when they use their phones to navigate between points in time, or what we call temporal mobility. Instead of staying in the moment that they have been allocated within a linear temporal framework, they travel through time to moments when they were, or will be, different—moments that display other facets of their personalities and other interests than what they are engaged in at present.

Furthermore, by anticipating many other such moments to come, people change their experiences of the present. In actions of media multitemporality, when past and future are juxtaposed onto the present, the present becomes more than a rushed moment between two meetings. From constraining, the present becomes open-ended: it is the future past that one would appreciate and feel nostalgic about in the future. With its newly achieved potential, the present is then better equipped to stand the test of time: it might be a fleeting moment, but it is not a moment to disregard. Acting on time is an ethical project (Moroşanu, 2016b), a way of expressing the values one wants to live by. Acceleration does not erase these values, as people find ways to salvage the present and to record evidence of lives that are being spent purposefully, and not always in a rush.

NOTE

This research was funded by the Research Council UK. The first study was funded by the Engineering and Physical Sciences Research Council of the UK (Grant no. EP/I000267/1). The second study was funded by the Economic and Social Research Council of the UK through the 'RCUK Innovation Fellowship fund' (Grant no. ES/R501104/1).

REFERENCES

Adam, B. 2004. *Time*. Cambridge: Polity Press.

Bear, L. 2014. "Doubt, Conflict, Mediation: The Anthropology of Modern Time." *Journal of the Royal Anthropological Institute* 20(1): 3–30.

Birth, K. 2012. *Objects of Time: How Things Shape Temporality*. New York: Palgrave Macmillan.

Castells, M. 1996. *The Rise of the Network Society*. Oxford: Blackwell.

Dourish, P. 2001. *Where the Action Is: The Foundations of Embodied Interaction*. Cambridge, MA: MIT Press.

Eriksen, T. 2001. *Tyranny of the Moment: Fast and Slow Time in the Information Age*. London: Pluto Press.

Glennie, P., and N. Thrift. 2009. *Shaping the Day: A History of Timekeeping in England and Wales 1300–1800*. Oxford: Oxford University Press.

Greenhouse, C. 1996. *A Moment's Notice: Time Politics across Cultures.* Ithaca, NY: Cornell University Press.

Gregg, M. 2018. *Counterproductive: Time Management in the Knowledge Economy.* Durham, NC: Duke University Press.

Guyer, J. 2007. "Prophecy and the Near Future: Thoughts on Macroeconomic, Evangelical, and Punctuated Time." *American Ethnologist* 34(3): 409–21.

Harvey, D. 1990. *The Condition of Postmodernity.* Oxford: Blackwell.

Horst, H., and D. Miller. 2006. *The Cell Phone: An Anthropology of Communication.* Oxford: Berg.

Keightley, E. 2013. "From Immediacy to Intermediacy: The Mediation of Lived Time." *Time & Society* 22(1): 55–75.

Madianou, M., and D. Miller. 2013. *Migration and New Media: Transnational Families and Polymedia.* London: Routledge.

Moroşanu, R. 2016a. *An Ethnography of Household Energy Demand in the UK: Everyday Temporalities of Digital Media Usage.* New York: Palgrave Macmillan.

———. 2016b. "Making Multitemporality with Houses: Time Trickery, Ethical Practice and Energy Demand in Postcolonial Britain." *The Cambridge Journal of Anthropology* 34(1): 113–24.

Moroşanu, R., and F. Ringel. 2016. "Time-Tricking: A General Introduction." In Special Section—Time-Tricking: Reconsidering Temporal Agency in Troubled Times. *The Cambridge Journal of Anthropology* 34(1): 17–21.

Mullan, K., and J. Wajcman. 2019. "Have Mobile Devices Changed Working Patterns in the 21st Century? A Time-Diary Analysis of Work Extension in the UK." *Work, Employment and Society* 33(1): 3–20.

Nowotny, H. 2005. *Time: The Modern and Postmodern Experience.* Cambridge: Polity Press.

Ofcom. 2014. *The Communications Market Report.* London: Ofcom.

Rosa, H. 2013. *Social Acceleration: A New Theory of Modernity.* New York: Columbia University Press.

Serres, M. 1982. *Hermes: Literature, Science, Philosophy*. Baltimore, MD: John Hopkins University Press.

Serres, M., and B. Latour. 1995. *Conversations on Science, Culture, and Time.* Ann Arbor: University of Michigan Press.

Sharma, S. 2014. *In the Meantime: Temporality and Cultural Politics*. Durham, NC: Duke University Press.

Tacchi, J. 2000. "The Need for Radio Theory in the Digital Age." *International Journal for Cultural Studies* 3(2): 289–98.

Urry, J. 2000. *Sociology Beyond Societies: Mobilities for the Twenty-First Century.* London: Routledge.

Virilio, P. 2012. *The Great Accelerator*. Cambridge: Polity Press.

Wajcman, J. 2015. *Pressed for Time: The Acceleration of Life in Digital Capitalism.* Chicago: The University of Chicago Press.

———. 2019. "The Digital Architecture of Time Management." *Science, Technology & Human Values* 44(2): 315–37.

Wajcman, J., and D. Dodd, eds. 2017. *The Sociology of Speed: Digital, Organizational and Social Temporalities.* Oxford: Oxford University Press.

Chapter Seven

Reconfiguring Synchronicity and Sequentiality in Online Interaction

Multicommuniciation on Facebook Messenger

Hannah Ditchfield and Peter Lunt

This chapter explores multicommunication within the context of the social networking site Facebook. Here, multicommunication is understood as instances where more than one thread of interaction occurs at the same time. Holding simultaneous and overlapping conversations in offline interactions is notoriously difficult to execute. However, certain technological affordances of social media platforms, such as conversational persistence and compartmentalization, make multicommunicating a more attainable practice within digital interactive spaces. Despite this, though, multicommunication online remains a challenging pursuit, requiring users to split attention between numerous interactive threads and prioritize some interactions over others. This chapter explores the temporal implications of this specifically in relation to (1) synchronicity and (2) sequentiality. How, for example, do users manage or negotiate time lapses in talk when more than one interaction is taking place and how do users decide what interactions to respond to first over others?

To address these concerns this chapter explores multicommunication from a microsociological perspective. By this, we mean that rather than attempting to theorise the macro changes digital communication has had on the way we communicate with one another and the way our relationship to time and place is changing, we approach time as something that is actively managed in everyday interactions. In this way, we treat social interaction as an accomplishment and deploy the temporal concepts of synchronicity and sequentiality to analyse how daily conversations and encounters are organized. From this perspective, this chapter is not concerned with how these configurations of time are represented, but rather in how these configurations are orientated to and managed within the micro details of text-based interaction. Specifically, we ask whether the material and architectural qualities of digital environments that afford practices such as multicommunication are altering the way temporal configurations of interaction are organized and whether such notions need reconsidering in the digital age.

To begin, this chapter outlines features of synchronicity and sequentiality that have been found to be useful in the analysis of face-to-face conversation and their potential relevance to online interaction. We then introduce the practice of multicommunication in more detail, and its specific relationship to Facebook Messenger, before moving on to discuss the theoretical framing of this research that relates temporally organized patterns of digitally mediated interaction to the technological and social affordances of social media environment. In this context, we go on to present an analysis of empirical examples of multicommunicating on social media. These examples are taken from Ditchfield's (2018) research on Facebook Messenger and are presented and analyzed to explore how temporal configurations are orientated to and managed in a social media landscape.

TEMPORALITY AND INTERACTION

Time and temporal configurations are at the heart of interactional organization and in face-to-face communication there are expectations around the length of gaps or pauses between interactional turns in face-to-face communication. Jefferson (1989), for example, argues that in certain cultures, the maximum tolerance of silence between turns in face-to-face contexts is around one second before there is the potential for interactional trouble to occur. These expectations are evident in the routine accomplishment of communication synchronicity in social interactions. A synchronous temporal order refers to instances when information is exchanged in a rapid fashion between interlocutors, the paradigmatic cases of which are face-to-face interactions and telephone calls (Madell and Muncer, 2007). Asynchronous orders refer to interactions where turn taking is much slower, for example, in letter writing (Madell and Muncer, 2007). Quasi-synchronicity refers to interactions that are cotemporal but not copresent in which the message construction process of the interaction is not available to interactional partners, for example, in instant messaging (Garcia and Jacobs, 1999). In addition to synchronicity, the timings between interactional turns, a second temporal concept, sequentiality, is integral to the organization of conversation and refers to the ordering of activities or turns in communication. Temporal orders in this sense can be successive (e.g., turns and activities occur one after another) or simultaneous (e.g., turns and activities overlap and occur at the same time). In face-to-face communication, the usual expectation is that interactional turns occur successively, in which individuals respond to utterances one at a time rather than talking over each other or responding at the same time. Our question in this chapter is whether, and to what extent, these temporal structures are reconfigured in the context of multicommunication and what this can tell us more broadly about the organization of time in our digital lives.

MULTICOMMUNICATION ONLINE

Research on multicommunication has often focused on how individuals manage simultaneous interactions across different sites. Licoppe (2004), for example, focuses on how overlapping interactions on Skype and face to face are negotiated and Reinsch, Turner, and Tinsley (2008) explore multicommunication in the work environment, specifically the overlap of online and offline work interactions. Here, though, our specific interest is in how several separate interactions are constructed within the *same* interactive space: that of Facebook.

Facebook as a social networking site offers users various modes of communication within the virtual walls of the single platform including writing status updates, commenting on others posts as well as less textual forms of interaction such as "shares" and "likes." Within this space, users also have the option to communicate via Facebook's own instant messaging service, Facebook Messenger. Facebook Messenger, like other instant messaging services such as Whatsapp, offers users the ability to send messages to specific groups or individuals inbox to inbox (rather than through profiles or newsfeeds). The difference between Facebook Messenger and Whatsapp, however, is that Facebook Messenger is situated within a broader platform context with other communicative functions available to users in the same virtual space.

In this way, multicommunication can be seen to occur in two ways on Facebook. The first is located solely within the function of Messenger with users holding multiple Messenger conversation threads at one time, switching from one to the next. The second is a form of multicommunication that occurs in the same site (Facebook) but between different modes of communication (e.g., Messenger interaction and comments). It is this second form of multicommunication that is specific to a platform like Facebook, rather than other popular messaging services. This chapter will analyze examples of both.

There are many material qualities of the Facebook platform, including the persistence and compartmentalization of text, that afford multicommunication in these two ways. Such affordances of social media have been linked to questions of privacy and surveillance and to the formation of networked publics (boyd, 2010; boyd and Ellison, 2007). Here, however, we are interested in how features of social media act as social and communicative affordances in which digital environments shape social interaction (Hutchby and Barnett, 2005; Hsieh, 2012).

There are parallels between the general affordances of social media of persistence, replicability, scalability, and searchability and the specific affordances of social interaction (boyd, 2010). Firstly, conversations on Facebook have the feature of creating a "persistent textual record of interaction" meaning that there is a typed record available to interlocutors to help them "keep track of what is going on" and reduce interactional incoherence (Herring, 1999). This allows for the possibility of multicommunicating as users can reread messages to remind themselves of what has previously been typed affording the potential to drop in and out of conversations returning to pick up the thread, a useful feature when more than one conversation is occurring at once. Secondly, interactions on Facebook are "compartmentalised," a feature that enables interactions with individuals to be separately contained (Reinsch et al., 2008). This feature enables Facebook chats, messages, and comments with different interlocutors within separate threads of communication. In the examples presented in this chapter, Facebook Messenger interactions were arranged in separate mini windows along the bottom of the computer screen so that the focal user of our analysis could monitor and respond to individual threads without the other participants being aware just as she/he was unaware of whether or not their interactional partners were engaged in other threads. In the face-to-face context, such compartments do not exist so that attempting to communicate in a multiple way, one could expect considerable overlap to occur within talk as well as considerable confusion as to which response was for whom. In this way then, compartmentalization of interaction threads online makes possible the management of synchronicity and sequencing in multicommunication.

These material qualities that make multicommunication possible are not unique to Facebook Messenger, or even instant message as a genre, as they are also affordances of written communication in general. One of the oldest communicative forms, that of letter writing for example, also provides writers, rather than typists, with a persistent textual record and separate compartmentalized interactions. Such opportunities for interaction then are not new. However, what is different about how these affordances manifest on Facebook Messenger is that they occur in an environment with more synchronous expectations than that of letter writing, or indeed emailing (Hutchby, 2014). In this way, the expectation and pressure to minimize time gaps and respond in a timely manner may be greater than in offline equivalents. In addition, there are alternative architectural features of Facebook Messenger interaction that are not present in letter writing. Features such as showing when users are "online" with a green (rather than grey) circle next to their name and indications within a thread of interaction whether a message has been "seen" both alter the temporal pressure that users may be under to respond within a shorter time frame.

The temporal ordering of messaging is further complicated in multicommunication as switching between threads has to be coordinated simultaneously with temporal ordering within threads. A material constraint arises because interactions are accomplished through a single keyboard/touchpad so attention can only be given to one thread at a time. Consequently, we expect that pauses within interactions will be extended but, in contrast to letter writing, not over long periods of time. Hutchby (2014) discusses these issues in relation to the way that email can be used both in a manner that reflects letter writing, with long pauses between posts or, in contrast, as a form of "texting" in which rapid turns create a sense of liveness. Our question is how do participants manage these constraints of sustaining a sense of copresence within interaction threads while managing the synchronicity within threads?

Multicommunication also raises challenges for the analysis of sequentiality online because of the relative independence of the different lines of communication or conversations. Multicommunication also requires sequential organization both within each interaction (e.g., the sequence and order of turns within a single thread or conversation) and sequentiality across interactions (e.g., the sequence and order interlocutors respond to different threads). In other words, the affordance of engaging in multiple interactions at the same time (synchronicity) raises some novel challenges to sequencing in balancing the sequencing of interaction within dyads while sharing attention and therefore sequencing participation across dyads. Our question is in what ways is this complex web of sequencing orientated to and managed in practice?

MULTICOMMUNICATION IN ACTION

To address the questions posed by this chapter, we present an analysis of material drawn from Ditchfield's (2018) research, which uses screen capture technology to record Facebook user's interactions. Most research exploring the phenomenon of multitasking and multicommunicating has tended to adopt a "broader, individual and cognitive perspective (. . .) largely omitting the detailed practices through which multiple activities are actually managed together, in real time, in social interaction" (Haddington, Keisanen, Mondada, and Nevile, 2014, 5). Such studies have approached the phenomenon as observed within face-to-face interactions and semimediated contexts using methods such as surveys, time logs, user diaries, and interviews (Reinsch et al., 2008; Junco,

2012; Judd, 2014). Such research is unable to address the questions we have raised about the temporal dimensions of the ordering of digitally mediated social interactions through synchronicity and sequencing.

To address this, we analyze the results of the application of screen capture records of moment-by-moment actions on Facebook, providing access to the "detailed practices" of multicommunicating. Such detailed recording of practices enables the measurement of response times and time lapses between interactional turns and an analysis of how users shift their attention or focus between simultaneous yet independent threads of communication. In our analysis, we draw on conversation analysis (CA) (Sacks, Schegloff, and Jefferson, 1974). At its most basic level "CA is the study of talk-in-interaction" and is the "systematic analysis of the kinds of talk produced in everyday naturally occurring situations of social interaction" (Hutchby, 2001, 55). A concern of CA is sequential organization, which here is applied to the temporal organization of Facebook Messenger interactions in which participants manage multiple lines of interaction at the same time. The following sections present an analysis of synchronicity and sequencing in interactions involving an individual simultaneously engaging in separate chats with three contacts who communicate with him without access to his interactions with the other participants.

SYNCHRONICITY

In addressing the issue, the first the question of the temporal organization of mediated social interactions in multicommunicating on Facebook we begin by examining the time lapses between interactional turns in the two extracts (1 and 2) below. These interactions occurred at the same time with the recording participant (Jamie) attending to the conversation with Daniel (extract 1) and Olivia (extract 2) simultaneously. The time delays between turns are displayed in brackets.

Extract 1:

Jamie: up to much?

[5 seconds]

Daniel: no not really mate you?

[36 seconds]

Jamie: naaaaaa- just doing some work- footy has been called off now

[51 seconds]

Daniel: ahhhhh

Daniel: how come?

[31 seconds]

Jamie: wind and rain apparently

[7 seconds]

Daniel: ah

[1 minute 9 seconds]

Jamie: you taken that car for a run about yet?

[5 seconds]

Daniel: not yet im going to tomorrow now cos traffic is daft round here at mo
[6 seconds]

Jamie: you are a useless person

<u>Extract 2:</u>

Olivia: Ameila Grace on fb

[32 seconds]

Jamie: oh yeah shes pretty nice- why would she eat me alive aha?

[1 min 8 seconds]

Olivia: shes just extremely independent and I think your probably a bit too nice

Olivia: I mean that in a nice way haha

[44 seconds]

Jamie: Haha, im not too sure how to take it aha

[4 minutes 4 seconds]

Olivia: because of the job we do she kind of go's for manly men she would just have you whipped

Olivia: shes a really nice girl though

[1 minute 4 seconds]

Jamie: hahaha, I wouldn't want a relationship with her

[1 minute 26 seconds]

Olivia: hahaa why shes a babe

Olivia: are you a player nowadays is that what your telling me haha

[37 seconds]

Jamie: hahaha, im not a playa- just scared of commitment lol

Jamie: need to find the right person *sick face*

The time lapses within these two examples display both synchronous and asynchronous temporal patterns. In extract 1, for example, more synchronous gaps of 31 and 7 seconds are apparent followed by longer, more asynchronous, gaps of over 1 minute. The temporal frame then returns to a quicker pace of 5 and 6 second gaps. Similarly, in extract 2, shorter synchronous gaps of 1 minute 8 seconds and 44 seconds proceed a lengthy asynchronous gap of over 4 minutes before returning to shorter gaps of 1 minute 4 seconds and 1 minute 26 seconds. It is apparent that the Facebook message interactions shown here can neither be described as simply synchronous or simply asynchronous. Instead, users move between these temporal frames fluidly, orientating to them in an interchangeable manner. In this way, neither synchronous nor asynchronous frames are applicable. Quasi-synchronous interaction, understood as interactions that are synchronous but where the message construction process of the interaction is only available to the speaker (or typer), is also not seen here. Rather an alternative quasi-synchronous pattern is apparent, a pattern that involves switching between different temporal paces meaning that the interaction is in some ways synchronous and in some ways not: it is seemingly both, yet neither.

A notable feature of this fluid temporal arrangement is that the longer time lapses in interaction are not made accountable by users, which we would expect if the usual constraints of temporal ordering of synchronous exchanges were in operation. The notion of accountability is an important one in work drawing on conversation analysis stemming from the ethnomethodological foundations of the approach as understanding individuals as "members" of a "collective" that share a set of anticipations, expectations or interactional rules (Bischoping and Gazso, 2016). These "rules" of interaction are seen to be used as a tool to help members "make sense of their environments of action" (Heritage, 1984, 292). Our actions within our conversations, then, are held accountable to these rules and we routinely signal our compliance with such rules as part of our interactions. In extracts 1 and 2, it can be seen that the time gaps between turns are considerably longer than the "acceptable" one second gap commonly seen within face-to-face or telephone interactions, thus deviating from the interactional rule originally highlighted by Jefferson (1989): that time lapses should be approximately one second.

However, in the data shown above even interactional gaps of over four minutes long are not orientated to or made accountable by users. Conversation instead continues in an unproblematic fashion despite the length of gap between turns. Due to the nature of the data collected here, it is not possible to make claims about how the participant feels about this gap or exactly why it was not made accountable. What can be argued though is that within the course of interaction itself participants do not make gaps relevant. In this way, we argue, such lapses are naturalized or domesticated. However, within the data collected for this research, there were exceptions to this general pattern in which the temporal pace of conversation became relevant.

Extract 3a:

Jamie: No you won't

Jamie: you'll have half a cider and give up [53 seconds]

Daniel: never

Daniel: dont drink cider [22 seconds]

Jamie: OK, a Heineken? [10 seconds]

Daniel: 5

[27 seconds]

Jamie: and then that's you done?

Jamie: thats what you call getting really drunk? [1 minute 36 seconds]

Jamie: ???? [34 seconds]

Daniel: nahhhh

Daniel: ill drink more

Here, what is of interest is the turn taken by Jamie on line nine. In a conversation where Jamie is questioning Daniel on what he will drink on an upcoming night out, Jamie asks "and then that's you done? that's what you call getting really drunk?" (lines seven and eight), which is followed by a 1 minute and 36 second time gap (the longest seen in this interaction at this point). Jamie then sends a further message of "????" (line nine). What occurs in this extract then is a shift between forms of communication synchronicity. The conversation begins by following a more synchronous response time pattern before dropping into a more asynchronous response time with a gap of over one minute. This fluidity between temporal patterns, though, does not go "unnoticed" and is not "naturalized" as it was in the previous examples because Jamie follows up these turns with a row of question marks. If the question marks were sent immediately after entries 7 and 8 it would have indicated that Jamie was simply eager to receive a response from Daniel. However, because these question marks were sent over a minute and a half after the questions asked, the question marks work to probe and encourage an answer following the gap in silence, thus making the increased time gap relevant and accountable to the interaction.

This example indicates that switching between synchronous and asynchronous patterns can be problematic and that the uncertainty around whether an interaction is synchronous or not is sometimes something that needs to be "worked out" within the run of interaction itself. Here, the "????" works to make visible and therefore accountable the situation (in terms of time frame) that the interactants are operating within. Later within the same interaction, an even longer interactional gap occurs:

Extract 3b:

Jamie: Haha why will you? What makes the 21st any different to every other night out?

[9 seconds]

Daniel: dunno

Daniel: just feel like it Daniel: first night back in country

[18 seconds]

Jamie: haha, have you been drinking whilst over there? [5 mins 36 seconds]

Daniel: nope

Daniel: no point

[17 seconds]

Jamie: looooool, your gonna be such a light weight

Here, a few turns on from the conversation seen within extract 3a, a longer time gap of over 5 minutes (5 minutes 36 seconds) occurs. Yet, this longer time gap, one that is significantly longer than the 1 minute 36 seconds previously discussed, is not made relevant in the interaction and is, similarly to extracts 1 and 2, "naturalized." This can be explained by the context of the interaction itself. See below the interaction that occurs prior to the time gaps of 1.36 minutes in extract 3a and 5.36 minutes in 3b:

3a:

Jamie: you'll have half a cider and give up Daniel: never Daniel: don't drink cider

Jamie: OK, a Heineken?

Daniel: 5

Jamie: and then that's you done? Jamie: thats what you call getting really drunk? [1 minute 36 seconds]

3b:

Daniel: I'll get on it

Daniel: I will be in the mood Daniel: its basically all day drinking

Daniel: 1pm until early hours aha

Jamie: haha why will you? What makes this day different to every other night out lol Daniel: dunno

Daniel: just feel like it

Daniel: first night back in country

Jamie: haha, have you been drinking whilst over there?

[5 mins 36 seconds]

In extract 3a Jamie is teasing Daniel about how much they can drink or will drink on an upcoming night out, suggesting that Daniel's "5" Heinekens is not enough. Due to the interrogative nature of the questions being asked before the interactive gap, it is a possibility that Daniel could take offense at the teasing propositions being put forward by Jamie, thus leading to interactional unrest. By interactional unrest, we are not referring to interactional trouble in relation to the coordination of interaction, but to the potential trouble a particular conversational topic brings. With an expanding time gap, the chance that Daniel would have taken offense to the questions asked becomes a more likely outcome, thus potentially probing Jamie to work to encourage a response (with "????") to help determine whether there is, indeed, any interactional unrest to resolve. In contrast, the nature of the question asked before the interactive gap in 3b is much less interrogative, with Jamie simply asking whether Daniel had been drinking whilst "over there" (the participant had been travelling for a few months prior to this conversation). Given that there is less of a risk of causing offence with this utterance, there may have been less of a need to "chase up" or encourage a response despite the significantly longer time gap of 5 minutes 36.

From the extracts presented, it is evident that when multicommunicating, users switch from more synchronous to asynchronous patterns. There are phases of the interactions that resemble live texting alongside phases that embody the slower pace of letter writing. Across the course of interaction, users slip between these two modes in an unaccountable fashion. This tells us that in instances of multicommunication synchronicity, and asynchronicity, are relevant concepts for the analysis of online temporal configurations. Users' manage synchronicity by switching between these two patterns during the interaction. Switching frames within interaction is not something new or uniquely applicable to online interaction. At dinner parties, for instance, individuals can switch between group discussions involving all to subgroup interactions between pairs or one end of the table and the other. What is seen here though is not a switch between who is involved in the interaction but instead the temporal frame that is adopted. As illustrated in extracts 3a and 3b, what temporal frame that is drawn upon can depend on the interactional context (e.g., the topic of conversation being discussed). Some interactional contexts, for example, are orientated by participants as requiring a synchronous frame, whilst for others, asynchronous time lapses suffice.

SEQUENTIALITY

Multicommunication requires sequential organization in two ways: (1) within single interactions (e.g., the way users' coordinate turns in their conversation) and (2) across interactions (e.g., the way the multicommunicating user moves between the separate, overlapping threads of communication). This discussion is specifically interested in exploring the latter: how users orient to sequentiality *across* multiple conversations. Sequentiality in this sense can be ordered in two ways. Firstly, sequential orders can be successive, in which activities occur one after the other. Secondly, sequential orders can be simultaneous, in which activities overlap to occur at the same time.

Mondada (2014) argues that whether temporal orders are successive or simultaneous is intrinsically linked to the resources available to an individual. For example, surgeons operate on patients whilst recording the procedure as a demonstration for teaching purposes as a form of simultaneous multiactivity in which participants engage in both oral (speaking about the operation) and physical (conducting the operation) activities at the same time. The resources used in this example are complimentary as the surgeons can utilize their voices and embodied movements in a combined fashion. In contrast, when engaged in multiple interactions on Facebook, the activities require the *same* material resources: one mouse, touchpad, or keyboard. Mondada argues that as soon as the same resources are required for two different courses of action "the participants have to switch from a simultaneous mode to a successive mode" (2014, 38).

In this way, then, multicommunication is technically organized successively as users cannot physically respond to more than one message at the same time. However, multiple conversations are open within the same time frame when multicommunicating users do not close one interaction before opening another. Instead, they keep numerous threads live at one time, moving between them to respond or initiate turns. In this sense interactions are simultaneously active, yet due to resource constraints, are successively attended to. In contrast to sequentiality, then, users do not switch between different temporal configurations (e.g., switching between synchronous and asynchronous). Instead, the temporal frames of sequentiality and simultaneity are overlapped and embedded within each other.

Despite recognizing the unique temporal arrangement here, questions remain around how individuals organize responding to their multiple interlocutors: How do they decide who to respond to first? How do they orient towards and manage these overlapping conversations? This chapter identifies two ways in which this is done: (1) through an embedded order and (2) through an exclusive order. The following extracts show two typical examples of response patterns seen within Ditchfield's (2018) data set. By response patterns, we are referring to the order in which Facebook users attend to the multiple interactions that they are engaged in. Extracts 4 and 5 show the order of activities that happen on user's screens, documenting how they move between their multiple interactions:

<u>Extract 4: Jamie</u>

→1. Receives message from Daniel
→2. Receives message from Sarah
→3. Receives message from Olivia
→4. Responds to Daniel
→5. Responds to Sarah
→6. Responds to Olivia

<u>Extract 5: Jamie</u>

→1. Receives message from Daniel
→2. Responds to Daniel
→3. Receives message from Olivia
→4. Receives message from Sarah
→5. Responds to Olivia
→6. Responds to Sarah

In the first example (extract 4), Jamie receives three messages in separate threads of conversation from three separate interactional partners: first Daniel, followed by Sarah, followed by Olivia. In this scenario, Jamie has three messages to respond to and is faced with a decision about who to respond to first. Quite simply, Jamie responds in the order in which he received the messages, replying to Daniel, followed by Sarah, followed by Olivia. In extract 5 Jamie is faced with a similar decision. Again, Jamie chooses to respond to the messages in the order in which they were received, first responding to Olivia followed by Sarah. These extracts reveal what Mondada terms an "embedded order" of multiactivity: an order that is "organised in an intertwined and alternating way" (2014, 35). The interactions above overlap in time (and are thus intertwined in terms of temporal frame) and are also orientated to in an alternating fashion. Specifically, these extracts are alternated on the basis of a "first come-first serve" pattern of response (e.g., whoever sends the message first is the first person that the Facebook user responds to and whoever sends a message second is responded to second).

In the data collected, this embedded approach to sequentiality was the most commonly adhered to. However, exceptions occurred as seen below:

Extract 6: Mark

1. Opens chat with Laura
2. Begins message to Laura
3. Comment notification from Josh
4. Clicks on comment notification
5. Likes comment
6. Continues with message to Laura
7. Sends message to Laura

In this extract, the recording participant Mark opens a chat and begins constructing a message to Laura (lines 1–2). Mark then receives a comment notification, which appears as a box in the top left-hand corner of the screen informing him that Josh has commented on a recent interaction between them. Mark clicks on this notification box, sending him through to the comment interaction thread and leaving the page in which he was writing the message to Laura (line 4). Mark then engages with Josh's comment through "liking" it before returning to the message with Laura and pressing send (lines 10–11).

There are two important aspects to highlight here. First, this is a form of multicommunication that occurs within the same *site* (Facebook) but across different *modes* (Facebook Messenger and Facebook comments). It is thus an example of multicommunication that is specific to the Facebook platform rather than other instant messaging services. It is also an example of a different form of response pattern. Rather than the "embedded order" that was highlighted within extracts 4 and 5, what is observed here is that of an "exclusive order"; one where "one activity is momentarily abandoned in order to carry out another" (Mondada, 2014, 35). What is seen is an occurrence where one activity, or interaction, is put on hold (in this case the construction of the message to Laura) to attend to another interaction (in this case the comment notification from Josh). Exclusive orders are not unique to online or multicommunication interactions. Offline, for instance, individuals can put on hold a face-to-face conversation to answer a phone call or pause a Skype chat to answer the door (Licoppe and Tuncer, 2014). However, what is different here is the way that neither interlocutors are aware of the competing interaction. In the case of extract 6, it is only Mark who knows he is balancing and putting on hold certain conversations—Laura and Josh are none the wiser.

Putting activities "on hold" is argued to demonstrate a level of hierarchy between the multiple interactions that are occurring at one time (Mondada, 2014). By suspending one activity to attend to another, participants reveal how one activity may be prioritised over another. In other words, participants actively display which activity is main and which activity is side (Mondada, 2014). As discussed by Goffman, the main involvement can be understood as the one that "absorbs the major part of an individual's attention and interest" and the side being the one carried out "in an abstracted fashion without threatening or confusing simultaneous maintenance of the main involvement" (Goffman, 1963, 43). In extract 6 the interaction with Josh via comment is prioritized over the interaction with Laura; the latter being "suspended" by Mark (line 4). Through this response pattern, Mark displays how his activity with Josh is his "main" involvement with the thread with Laura taking the position at the "side." This reflects Goffman's (1961) account of complex social encounters with groups of people playing different communicative roles organized so that aspects of engagement are treated as focal and others as peripheral in a dynamic fashion across the encounter.

One difference between these two overlapping interactions is the sense of co-presence within the conversations. Josh, for example, shows signs of being virtually present on Facebook as he has, in real time, responded to one of Mark's photos thus producing a live notification. In the chat with Laura, however, Mark is *opening* the conversation meaning that Laura is not yet (virtually) present or engaged within the interaction. In fact, Laura is unaware that this interaction is about to take place. Extract 6 then highlights how more synchronous interactions, or interactions where the presence of the interactional partner is more pronounced, are given greater precedence over interactions in which interactional engagement is yet to begin. In this way, what pattern of sequentiality is followed, be it embedded or exclusive, can depend on the interactional context of the multiple threads engaged in. However, unlike in relation to synchronicity where conversational topic influenced temporal configurations, here it is the perceived level of copresence.

CONCLUSION

This chapter has explored the temporal implications of multicommunication on both synchronicity and sequentiality. Through the analysis of Facebook Messenger data, we have revealed how user's switch between more synchronous and asynchronous patterns during the course of interaction and how sequential and successive modes of communication are overlapping in nature. For both synchronicity and sequentiality, interactional context plays a consequential role in relation to which temporal configuration takes priority. Conversational topics, for instance, can impact whether more synchronous or asynchronous speeds of interaction are more suited and indicators of copresence can influence whether embedded or exclusive orders of response are orientated to.

In these ways the material qualities of digital environments that afford the possibility of multicommunication are altering temporal organizations of online interactions. Specifically, what are usually understood as binary temporal concepts in face-to-face communications are instead orientated to in a hybrid way in which synchronous and asynchronous and successive and simultaneous configurations are combined. The interactions analyzed in this chapter are not easily defined as simply "synchronous" or "asynchronous" or "successive" or "simultaneous." Instead, such dichotomies are merged with users fluidly switching between, or overlapping, these usually distinct modes. Of course, there are instances in offline communication where switching and overlapping of temporal configurations also occurs. Take the switches that occur between group and subgroup conversations at a dinner table or the way a phone call will be "put on hold" to attend to a summons. However, due to affordances such as textual persistence and compartmentalization that make multicommunication possible on platforms like Facebook, these more hybrid forms of temporal arrangements are arguably becoming more common place.

To further understand the impacts of multicommunication and hybrid temporal configurations, future research could reach beyond the discursive perspective to understand the perspectives of social media user's themselves. How do they experience multicommuication online and in what ways to they consider temporal organizations? Beyond temporal concerns there are questions to ask here in relation to interactional order, for example, do these changes in temporal organization have impacts on the so-called quality of interaction and the presentation of a moral self online.

REFERENCES

Bischoping, Katherine, and Amber Gazso. *Analyzing talk in the social sciences: Narrative, conversation and discourse strategies*. London: Sage Publications Ltd, 2016.

boyd, Danah. "Social Network Sites as Networked Publics: Affordances, Dynamics, and Implications." In *Networked self: Identity, community, and culture on social network sites,* edited by Zizi Papacharissi, 39–58. New York: Routledge, 2010.

boyd, Danah, and Nicole Ellison. "Social Network Sites: Definition, History and Scholarship," *Journal of Computer-Mediated Communication*, 13 no.1 (2007): 210–30.

Ditchfield, Hannah. "*Behind the screen of Facebook: An interactional study of pre-post editing and multicommunication in online social interaction.*" PhD Diss., University of Leicester, 2018.

Garcia, Angela, and Jenifer Baker Jacobs. "The eyes of the beholder: Understanding the turn-taking system in quasi-synchronous computer-mediated communication," *Research on Language and Social Interaction,* 32 no.4 (1999): 337–67.

Goffman, Erving. *Encounters*. Indianapolis: Bobbs-Merrill, 1961.

———. *Behaviour in public places: Notes on the social order of gatherings*. New York: The Free Press, 1963.

Haddington, Pentti, Tiina Keisanen, Lorenza Mondada, and Maurice Nevile. "Towards multiactivity as a social and interactional phenomenon." In *Multiactivity in social interaction: Beyond multitasking*, edited by Pentti Haddington, Tiina Keisanen, Lorenza Mondada, and Maurice Nevile (Amsterdam: John Benjamins Publishing Company, 2014), 3–32.

Heritage, John. *Garfinkel and ethnomethodology*. Oxford: Basil Blackwell, 1984.

Herring, Susan. "Interactional coherence in CMC," *Journal of Computer–Mediated Communication*, 4, no.4 (1999).

Hsieh, Yuli. "Online social networking skills: The social affordances approach to digital inequality," *First Monday*, 17 no.4 (2012).

Hutchby, Ian. *Conversation and technology: From the telephone to the internet.* Cambridge: Polity Press, 2001.

———. "Communicative affordances and participation frameworks in mediated interaction," *Journal of Pragmatics*, 72 (2014): 86–89.

Hutchby, Ian, and Simone Barnett. "Aspects of the sequential organisation of mobile phone conversation," *Discourse Studies: An Interdisciplnary Journal for the Study of Text and Talk*, 7 no.2 (2005): 147–71.

Jefferson, Gail. "Preliminary notes on a possible metric which provides for a 'standard maximum' silence of approximately one second in conversation." In *Conversation: An interdisciplinary perspective,* edited by Derek Roger and Peter Bull, 166–96. Clevedon: Multilingual Matters Ltd, 1989.

Judd, Terry. "Making sense of multitasking: The role of Facebook," *Computers & Education*, 70 (2014): 194–202.

Junco, Reynol. "In-class multitasking and academic performance," *Computers in Human Behavior*, 28 no.6 (2012): 2236–43.

Licoppe, Christian. "'Connected' presence: The emergence of a new repertoire for managing social relationships in a changing communication technoscape," *Environment and Planning D: Society and Space,* 22 no.1 (2004): 135–56.

Licoppe, Christian, and Sylvaine Tuncer. "Attending to a summons and putting other activities 'on hold': Multiactivity as a recognisable interactional accomplishment." In *Multiactivity in social interaction: Beyond multitasking*, edited by Pentti Haddington, Tiina Keisanen, Lorenza Mondada and Maurice Nevile (Amsterdam: John Benjamins Publishing Company, 2014), 169–90.

Madell, Dominic, and Steven Muncer. "Control over social interactions: An important reason for young people's use of the internet and mobile phones for communication?" *Cyberpsychology & Behaviour*, 10 no.1 (2007): 137–40.

Mondada, Lorenza. "The temporal orders of multiactivity: Operating and demonstrating in the surgical theatre." In *Multiactivity in social interaction: Beyond multitasking*, edited by. Pentti Haddington, Tiina Keisanen, Lorenza Mondada and Maurice Nevile (Amsterdam: John Benjamins Publishing Company, 2014), 33–78.

Reinsch, Lamar N., Janine Warisse Turner, and Catherine H Tinsley. "Multicommunicating: A Practice Whose Time Has Come?" *Academy of Management Review*, 33 no.2 (2008): 391–403.

Sacks, Harvey, Emanuel Schegloff, and Gail Jefferson. "A simplest systematics for the organisation of turn-taking for conversation," *Language*, 50 no.4 (1974): 696–735.

Part III

MAKING TIME FOR . . . COMMODIFICATION

Chapter Eight

Move Slow and Contemplate Things

An App That Drops Users Out from Distracting Aspects of the Internet

Alex Beattie

INTRODUCTION

> Turn on, tune in, drop out.
>
> —Timothy Leary, 1999

> Disconnection is the new counterculture.
>
> —Nicholas Carr, 2010

In Silicon Valley, the world's most famous place of technological innovation, technology professionals are rejecting their own inventions and dropping out[1] from the internet. According to reports, programmers, software engineers and user experience (UX) designers believe their designs and algorithms "hijack" people's brains (Lewis, 2017). These technology workers are not disillusioned by digital technology per se, but rather by the ideological and socio-economic system underpinning digital technology. This system is the "attention economy," which at a micro level, incentivizes UX designers to create habitual technological experiences that glues users to their screens and, at a macro level, reorganizes the internet into a persuasive advertising platform (Wu, 2016). Not only are some technology workers disconnecting from their own creations but, in true Silicon Valley style, are also inventing technologies to protect their attention and *drop* themselves out from aspects of the internet they perceive as preventing them from acting upon their intentions.

A technological dropout from the internet suggests a resurging interest in the quality of mediated time. Maya Indira Ganesh (2018, para. 9) observes that previous technologies such as radio or television caused temporal disruptions for users; what may be different today is the degree that UX designers consider themselves responsible for temporal manipulation: "It is an oft-repeated observation that when email was first available it was something of a reprieve from having to respond immediately, as you might to a phone call. Technology made a promise that you could manipulate, expand and make time; but this delay has become perverted, and UX became a sort of handmaiden to this effort." The challenge for UX design is that the core industry goal of maximizing user engagement clashes with the "looming expectation that everyone must become an entrepreneur of time control" (Sharma, 2014, 138). As such, organizations such as the Center for Humane Technology (CHT) (formerly Time Well Spent) calls for new UX standards and incentives beyond creating user habits to encourage designs and interfaces that more broadly empower the user (CHT, 2019). As part of this movement, a loosely affiliated group of applications (apps) and practices have converged around the CHT, advocating for "humane" design or "digital well-being." The implication here is that the right mix of design principles and software protocols can curb excessive time on a device and encourage the user to partake in healthier behavior.

One app that combines the ideals of well-being and time is Siempo, a portmanteau of the Latin words *sempre* (always) and *tiempo* (time) to denote "a mindful use of one's time" (Beattie, interview with Dunn, 2018). This chapter interrogates how Siempo orients time for its users and questions the sociocultural and ideological implications of Siempo's structuring of time. My method of analyzing Siempo is the walkthrough method (Light, Burgess, and Duguay, 2016), supplemented by an interview with Siempo chief executive officer Andrew Dunn, undertaken in 2018. The walkthrough method derives from science and technology studies as well as cultural studies and seeks to understand how an app is expected to be used by identifying what software protocols and cultural practices are encoded into the interface to guide user behavior. My walkthrough of Siempo highlights features that combine behavioral design and the Silicon Valley ethos of "intentionality" that anticipate and curb unconscious smartphone use, enabling the user to "drop out" of social media and other online distractions.

Of course, Silicon Valley and UX designers are not the first to try and shape individuals' experiences of time. Michel Foucault (1977) explored how the clergy, military, and prisons transformed bodies into sites of power through the use of time and schedules: "Time penetrates the body and with it all the meticulous controls of power" (152). I contend that the technological dropout provided by Siempo modulates a qualitatively slower orientation to time: not only does Siempo encourage a reduction of time spent on a device but also instructs users on *how* to spend their time on a device. It is in the manner of how users are prompted to use their device that a qualitative type of temporality emerges, which I call *intentional slowness*. Intentional slowness is a type of measured temporality-by-design, which occurs when intentional or goal-orientated behavior is favored over reactionary or habitual actions. Intentional slowness is distinct from advocates of slow living (Craig and Parkins, 2006) or slow media (Rauch, 2018), which call for alternative and moderate approaches to using media. Instead, my conceptualization of intentional slowness and walkthrough of Siempo is closer to studies that seek to understand how an interface can shape temporal experience (Ash, 2015; Dieter and Gauthier, 2019); in particular, my analysis follows Wajcman's (2019, 15) observation that Silicon Valley sees time as a "a quest ripe for technical fixes."

What follows in this chapter consists of two parts: the first offers a brief history of dropout counterculture in Silicon Valley to identify the origins of intentional slowness and provide a cultural context to Siempo. Drawing upon Fred Turner's seminal text *From Counterculture to Cyberculture* (2006), I argue the practice of dropping out from mainstream society to create and partake in alternative social structures reinforced design, technology, and personal transformation as integral to addressing social issues. Today, dropping out lives on in the form of countercultural events such as Burning Man and the health practice of digital detoxes. In the second part I walk through Siempo to highlight expectations of use and how an intentionally slow use of the smartphone is subtly enforced. I identify features of intentional packaging and batched notifications that utilize counterculture and health ideals to encourage the user to aspirational ends and self-discipline in their digital activities. I conclude by suggesting intentional slowness is indicative of broader neoliberal politics where alternative lifestyle and consumptive practices are positioned as the ideal means to solve societal problems.

SILICON VALLEY PRACTICE OF DROPPING OUT

The practice of intentionality, or goal-orientated living, is an idea that is central to Silicon Valley dropout counterculture. There has been little discussion of intentionality in relation to the ethos of Silicon Valley and its technological inventions. Previous cultural analyses of Silicon Valley and time have suggested that the tempo set by the industry is built upon the hacker maxim "move fast and break things" (Taplin, 2017; Wajcman, 2019). In contrast, I believe elements of dropout counterculture are informing design practices and apps, whereby temporality is less determined by an ethos of speed and disorder and more by a counterculture that historically celebrated introspection and deliberate behavior.

Dropping Out From Mainstream Society: New Communalists and Burning Man

In the late 1960s, tens of thousands of Northern Californians retreated from society to live in alternative communities and seek social cohesion and optimal methods of living. In a cultural history of Silicon Valley, Fred Turner (2006) christened a loosely affiliated group of Californian dropouts who retreated from mainstream American society in the 1960s and 1970s as the New Communalists. New Communalists sought communal living in rural areas as a response to the increasing speed of consumption in the 1950s and the alienation of life during the Cold War. Many New Communalists were technologists, building high-tech geodesic domes in the style of Buckminster Fuller, and connected to reach other via informational networks such as Stewart Brand's magazine *Whole Earth Catalogue*. The *Whole Earth Catalogue* presented tools and technologies to help New Communalists live leisurely, self-reliant, and spiritual lifestyles. According to Turner (2006) the New Communalists were largely apolitical, which distinguished them from the New Left—another countercultural group that emerged in the 1960s. Both the New Communalists and the New Left sought transformative societal change, but while the New Left was a traditional political movement that published manifestos and organized rallies, the New Communalists turned "toward technology and the transformation of consciousness as the primary sources of social change" (Turner, 2006, 4). On a similar note, cultural critic Christopher Lasch (1979, 4) contends the New Communalists were fixated on "psychic self-improvement" rather than widespread socio-political change. Turner (2006) positions the New Communalists as the cultural backbone of Silicon Valley cyberculture, informing what he and others have labeled the "Californian ideology": the unwavering belief in the emancipatory potential of technology (Barbrook and Cameron, 1996).

Today's equivalent to New Communalism might be found in so-called intentional communities. Although intentionality is well discussed in academic fields such as technology philosophy and phenomenology, as a folk concept it is underscrutinized. In folk communities intentionality generally refers to goal-orientated behavior (Cushman, 2015), or conduct entailing belief, desire, intention, awareness, or skill (Malle and Knobe, 1997). According to Shenker (2010, 6), an intentional community is "a relatively small group of people who have created a whole way of life for the attainment of a certain set of goals." What distinguishes intentional communities from typical tribes, villages, or other clusters that spontaneously emerge over the years, is that an intentional community is a determined group of individuals who come together to create an entire way of life. Intentional communities aim to achieve cohesive social integration, with some communities using behavioral design to organize and create governance and social structures. Twin Oaks Community is one example of such a community; inspired by B. F. Skinner's *Walden Two* (1948), Twin Oaks used behaviorist techniques to design its social systems. Twin Oaks Community no longer uses behaviorism and now calls itself an intentional community, reinforcing the link between design practices and countercultural ideals of intentionality. Odell (2019) suggests communities such as Twin Oaks are ahistorical, largely insular from political change and difference, and can be interpreted as a form of "lifestyle libertarianism": the practice of self-governance that is free from outside interference.

Given the prevalence of libertarianism in the technology industry (Taplin, 2017), it is perhaps unsurprising that intentionality has crept into technology culture. In a profile of notable design ethicist and CHT leader Tristan Harris, Bosker (2016) reports that Harris lives in an "intentional community house" in San Francisco with other entrepreneurs. Instead of retreating to a commune away from society, Harris appears to have implemented ideas of intentionality into his home and everyday practices. Moreover, Burning Man, arguably the apotheosis of countercultural hedonism, is viewed by some attendees as an intentional event. Brian Doherty describes Burning Man as a "true experiment in intentional creative community" (qtd. in Steinhauer, 2018, para. 3), and there are stories of Burning Man patrons ("Burners") transitioning to more permanent intentional communities (UPLIFT, 2015). Throughout the week-long party, Burners integrate creativity and performance into local economies, share resources, and partake in conscious or deliberate behavior (Turner, 2009).

Events such as Burning Man situate counterculture as consumable to new audiences. Zandbergen (2012, 364) notes how Silicon Valley countercultural communities in the 1960s and 1970s endorsed a type of participatory consumerism that "liberated emancipated commodities from imposed conventions about their use." While Burning Man has typically been framed as rejecting conventional consumerism (Gilmore and Van Proyen, 2005), Kozinets (2002, 36) contends the rituals of exchanging gifts are attempts to reinvigorate "consumption into a liberatory pursuit distanced from market logics." The appeal of Burning Man is the contrast it provides to "conformist and alienating superficiality of media-saturated and consumerism-fuelled, corporate capitalist mainstream culture" (Gauthier, 2013, 144). What can be gleaned from these studies is that dropout culture does not wholly reject consumerism but rather attempts to disrupt consumerism from conventional market logics. As local and temporary as countercultural modes of consumption at Burning Man may be, such disruption broadens the scope for consumerism to capture new varieties of individuated authenticity and creativity. As McGuigan (2009) notes, countercultures serve an important function to late forms of capitalism by repackaging commodities to appeal to younger, idiosyncratic demographics.

Burning Man also has a specific appeal to technologists. According to Turner (2009), Google's founders, Larry Page and Sergey Brin, have attended Burning Man since 1999, and encourage Google team members to attend to find artistic inspiration for their work projects. To Google, Burning Man presents an untapped common of ideals, processes, projects, and artists that is beneficial to creative work (Turner, 2009). By meshing labor processes and culture in such a way, Burning Man offers a cultural framework for new media production. Turner writes,

> As once, 100 years ago, churches translated Max Weber's protestant ethic into a lived experience for congregations of industrial workers, so today Burning Man transforms the ideals and social structures of bohemian art worlds, their very particular ways of being "creative," into psychological, social and material resources for the workers of a new, supremely fluid world of post-industrial information work. (2009, 75–76)

Google workers at Burning Man and intentional community houses in San Francisco indicate a trend in Silicon Valley whereby an idea or practice is transplanted from its countercultural origins and instrumentalized for work or everyday practice. New Communalism, intentional communities, and Burning Man all reinforce the notion that dropping out is not merely a practice of disconnecting from mainstream society, but generative of new ways of working and living.

Dropping Out from Digital Society: Digital Detoxes and Calm Design

Today, people drop out from the digital milieu. Popular social media platforms such as Facebook, Twitter, and Instagram are accused of foregrounding antisocial culture and behavior; for example, Facebook is accused of spreading disinformation as well as perpetuating outrage and body shaming culture (Vaidhyanathan, 2018). A popular type of digital dropout is undergoing a period of abstinence from digital media technologies, otherwise known as digital detoxes. Digital detoxes are increasing popular activities; in the summer of 2016, more than fifteen million British internet users undertook a digital detox (Ofcom, 2016), and off-the-grid holidays are on the rise (Cartwright, 2018). Digital detoxes have links to previous iterations of dropout culture—for example, the digital detox summer retreat Camp Grounded in Texas was created by former Burner Levi Felix (Fish, 2017). In an ethnographic study of Camp Grounded, Fish (2017) suggests digital detox camps merge the principles of New Communalism with postdotcom entrepreneurism and eco-spiritualism of the 2000s, all of which are anchored in the individual pursuit of wellness.

A digital detox frames a digital dropout as a health-related activity, with social media negatively associated as unhealthy or providing a superficial experience (Sutton, 2017). The utilization of a health frame reinforces a neoliberal ideology that Fish (2017) calls "digital healthism," which positions the individual as responsible for managing their own digital consumption. Digital healthism suggests how digital-free retreats have simultaneously become a luxury or "corporate amenity like a gym or a cafeteria" (Odell, 2019, 33), as well as a public health initiative. Digital detox campaigns like Scroll Free September (Royal Society for Public Health, 2018) and Pause (Capsana, 2019), frame digital dropouts as activities individuals can undertake to better manage their digital habits. When outlined as a public health issue, a digital detox loses some of the original counterculture lustre in exchange for normative legitimacy. Scroll Free September and Pause still reinforce digital healthism by encouraging people to undertake self-care practices of disconnection but aim to make such practices accessible to all interested citizens.

Similar to the New Communalists, modern-day Silicon Valley dropout culture does not reject media technologies per se, but instead rejects how media technologies are conventionally used, and should be designed. Ambient or minimalist ideals are influencing the design of technologies and attempt to shift technology use away from the mainstream context of the attention economy. Instilling digital media interfaces with a contemplative spirit is the aim of calm technology (Weiser and Seely Brown, 1995) or calm design (Case, 2015) and a subfield of human-computer interaction (HCI) called "positive computing"—a combination of HCI and positive psychology (Calvo and Peters, 2014). Calm technology or design, positive computing, and organizations such as the CHT are the latest incubators of the deep-seated Silicon Valley cultural belief that alternative lifestyle practices and revamped technology and design can address social issues and maximise human potential (Zandbergen, 2012).

The Intentional Slowness of Siempo

An example of a technology that addresses the issue of the distracting internet is an Android app and Chrome browser extension called Siempo. The purpose of Siempo is to disconnect or drop the end user out from common features of the smartphone that are distracting or encourage mindless phone usage. Features of the smartphone that are considered by Siempo to be potentially distracting are the interface and notifications. Siempo utilizes the countercultural ideal of intentionality to reorganize these features and qualitatively change the user's experience of time.

Intentional Packaging of the Interface

To encourage the user to undertake intentional or goal-orientated online behavior, Siempo comprehensively reorganizes and redraws the Android smartphone interface (figure 8.1). Once the user installs the app they must provide consent to allow Siempo to overhaul the standard Android home screen.[2] If consent is given, all colors and app branding of the interface are stripped away, replaced by a pallid grey background. The home screen is left largely blank except for a question field prompting users to consider their intention. The white space and decluttered interface of Siempo echoes the minimalist principles of calm design (Case, 2015) and is the basis of an inconspicuous aesthetic that Portwood-Stacer (2012a) argues can influence taste in media and, in turn, drive media rejection.

The aim of the Siempo interface is to disincentivize smartphone usage, which draws similarities to the goal of plain packaged cigarette packets. In New Zealand and Australia, cigarette packets are packaged with uniform and dull aesthetics to reduce the appeal of smoking. Addiction experts justify plain packaging on the basis that smoking is unhealthy and the use of branding, colors, and imagery on cigarette packets increase the appeal of smoking (Dewe, Ogden, and Coyle, 2013; Wakefield et al., 2013). The plainness of the Siempo interface draws upon similar findings in behavioral psychology to make the smartphone less appealing. But unlike plain cigarette packets, Siempo does not aim to prevent a behavior in general but, more specifically, encourages the user to be intentional with the way they use their phone. Users are prompted to set their intention via an open field text box (figure 8.1). The box asks: "What's your intention?" Some examples are provided and include: "spend more time with family," "eat healthy foods," and "keep my phone locked." Once users have set their intention, this text stands out on the screen.

If the users wish, they can circumvent the plainness of the intentional interface and customize the background image. There are no restrictions on what images can be chosen; if the user wants to, they could select a lurid image or a picture that promotes an unsocial or, ironically, prosocial media behavior, for their background. The implication is that the plainness of the interface is not of primary importance; the bigger priority is that the user sets an intention. Andrew Dunn explained to me that the purpose of the intention prompt is to gently nudge the end user to reflect on their smartphone use:

> I think that having an intention—whether it's going to a meeting or waking up in the morning for this time of the year—is a really way to invite in what you're

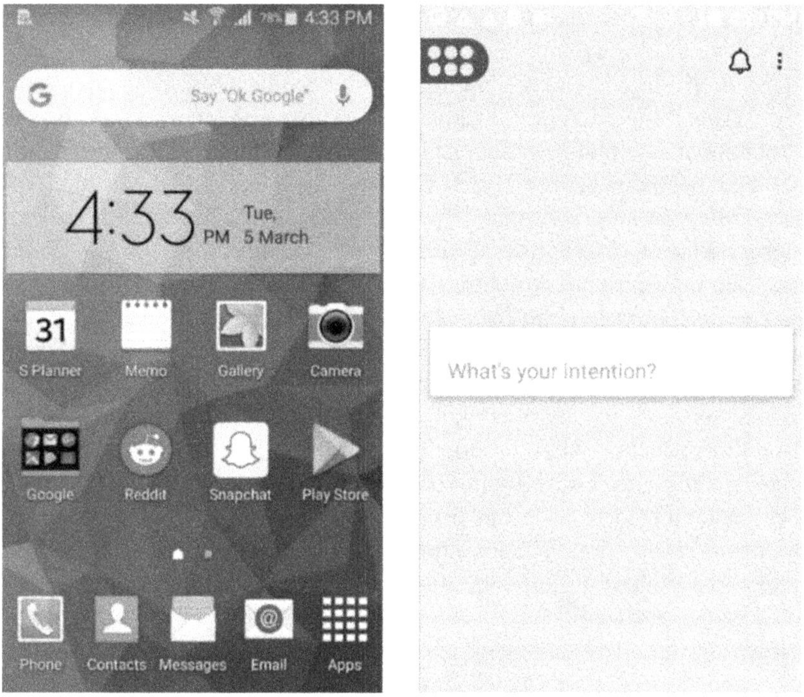

Figure 8.1. The intentional packaging of Siempo

> looking for in life, and to keep yourself on the track you want to go on. [Having an intention is about] what you want to focus on, umm there's . . . I dunno . . . it's kind of a loose subject like there's no real science behind it right now so we're just playing with different ways of surfacing it.

Andrew's comments reinforce that the benefits of setting an intention are not based on empirical evidence, distinguishing Siempo from the plain packaging of cigarettes. There is scientific evidence that cigarette smoking is harmful (Das, 2003), and in New Zealand and Australia the plain packaging of cigarettes is just one public health policy intervention alongside policies of price hikes and the hiding of products that share the goal of reducing smoking uptake. Plain packaging is a regulatory intervention with clear policy objectives grounded in empirical evidence that smoking cigarettes is harmful (Dewe, Ogden, and Coyle, 2013). More broadly, the field of health psychology justifies manipulating people to partake in intentional behavior in situations to address gambling or obesity (Ohtomo, 2013). In contrast, the debate about whether social media or digital screens are harmful is highly contentious. A number of studies dispute any causality between screens, social media, and poor mental health (Orben, Dienlin, and Przybylski, 2019; Reer, Tang, and

Quandt, 2019). Any claims by Siempo that their intentional interface will improve well-being are not grounded in empirical evidence.

Instead, Andrew's belief that there are benefits to setting an intention derives from Silicon Valley dropout culture. Siempo draws upon the ethos of intentional communities and Burning Man to encode the ideal of goal-orientated behavior within the smartphone interface. Whereas alternative communities became intentional by rejecting conventional social structures to design their own systems and rules based on shared social values, Siempo rejects the conventional structure of the smartphone interface in favor of a rigid design that insists the individual user set an intention and act upon it. Users are reminded of their intention every time they unlock their phone, as well as when they swipe to additional screens. The creation of an intentional interface by Siempo suggests that Silicon Valley counterculture does not only inform Silicon Valley production processes (Turner, 2009) but also the design of interfaces. Users are not entirely disconnected from the internet but instead encouraged to drop out from any distractions that could deter them from their set intention. The intentional interface of Siempo turns the smartphone into a kind of digital commune-for-one where a focused and purposeful life can be realized.

However, the intentional packaging of the Siempo interface appears to fall foul of the "designer's fallacy"—a deistic belief that the designer has God-like power in designing into a technology all intended purposes and usages (Idhe, 2008). It is doubtful that any user would want to use their smartphone intentionally all the time; yet there appears to be a presumption by Siempo that the user should always have an intention, an unlikely scenario given intentionality is only one component that motivates behavior alongside willingness and habit (Ohtomo, 2013). By frequently requesting the user set their intention, Siempo encourages purposeful behavior and risks foreclosing any spontaneous or unintentional usages of the smartphone. It is in how Siempo encourages users to plan and manage their smartphone usage that: "Power is articulated directly onto time, it assures its control and guarantees its use" (Foucault, 1977, 160). Moreover, the effects of constant goal setting via Siempo is the normalization of self-governance, similar to other smartphone monitoring apps that impose "endless micro-project management, transforming downtime into something structured, obedient, and explicitly purposeful" (Beattie, 2018). On Siempo there is less room for idle smartphone play or digital wayfaring; the implication to users is to actively manage all activities as part of a wider intention. As all human behavior cannot be intentional, the actual effect of the Siempo interface is to nudge users to self-manage their digital consumption and aspire to healthier, productive, or otherwise self-optimal modes of living.

Batched Notifications

A second feature of Siempo batches notifications at temporal intervals. Users can choose when they wish to receive notifications via a bell icon on the top right-hand side of the home screen (figure 8.1). When selecting the bell icon, the user is taken to an additional screen where they are asked when they want to receive notifications (figure 8.2). There are three options to batch notifications: as soon as they arrive (the default option on most phones); at customizable blocks of time; or only at a certain period of the day. To achieve either the second or third option, Siempo utilizes the Do Not Disturb functionality

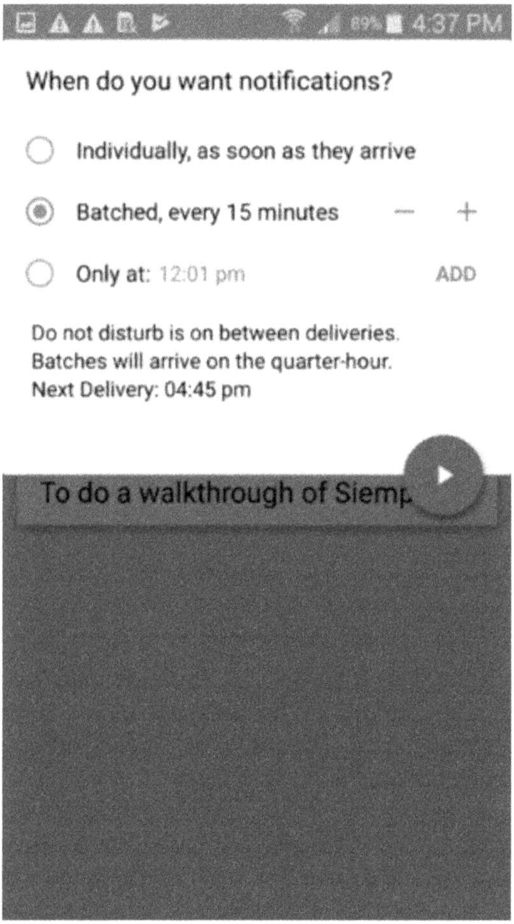

Figure 8.2. Batched notifications

on Android devices to silence notifications until the user had indicated when they want to receive them.

By delivering notifications on a schedule of the users' choice, Siempo embeds tactics of delay to intermediate a slower media experience. Users are not fully disconnected from the internet; rather, they are "dropped out" from the subjective uncertainty that is set by the irregular arrival of communications via notifications. From an informatics or information engineering perspective, a delay on the standard notification delivery process is an inefficiency but, from a Silicon Valley dropout perspective, the delay offers greater autonomy to perform uninterrupted intentional activities. As users can determine when they are interrupted, batched notifications offer a micro-temporality or an alternative schedule for the user. Being able to stagger notifications also directly relates to the name Siempo; as Andrew explains: "*Siempo* refers to always being mindful of your time and it also invokes simple and tempo, living your life at your own tempo" (Beattie, interview with Dunn, 2018).

The practice of delaying notifications also results in a more modest digital experience. Susanna Paasonen (2016) argues the difference between an unwelcome and a welcome digital distraction is impossible to categorize and depends on the user's subjective position. As Siempo bundles all notifications into the category of distractions, what batched notifications deliver is not just a slower media experience, but a dampening of what Paasonen (2018) calls "affective intensities"—the wide range of emotive feelings users commonly experience when connected to the internet. Affective intensities include the elation of receiving a congratulatory email, the boredom of scrolling social media, or the shame associated with receiving little to no social media responses. By temporarily depriving the opportunity for affective connection, batching notifications is akin to practicing asceticism, that is, a momentary opt-out of the roller-coaster of online emotions in favor of periods of concentration and purpose.

Asceticism is often something an individual chooses to undertake in the interest of self-improvement (Portwood-Stacer, 2012b). Delaying notifications is therefore a reminder of Foucault's (1997, 282) interpretation of asceticism as, "an exercise of the self on the self by which one attempts to develop and transform oneself, and to attain to a certain mode of being." As I have argued in this chapter, a lauded mode of being in the Silicon Valley is intentionality, with Siempo structuring time as an instrument for the user to manipulate in the wider goal of managing themselves as autonomous and self-regulated but, critically, productive agents. Unconscious usages of the smartphone are implied as unhealthy or inappropriate for an increasingly time precious society. To this end, the Siempo regime to self-govern behavior draws similarities to the latest means of efficiency engineering (Gregg, 2018), or work management and/or productivity tools (Guyard and Kaun, 2018).

CONCLUSION

In a Siempo (2019) video to potential investors, Andrew Dunn invites viewers to take deep breaths and reflect about the anxiety that can come from "mindless consumption" and technology use. Dunn describes Siempo as an intermediary technology that is "on your side" to assist users to make intentional technology and consumptive choices (Siempo, 2019). The video is a reminder that Silicon Valley's drop-out ethos lives on in Siempo. Dunn does not deem consumption per se as problematic; rather, it is *conventional* consumption, which in the digital milieu means mindless scrolling, which is the problem. Siempo channels the ethos of dropout counterculture into an intentional interface that intermediates the relationship between the user and their unconscious digital habits. Alongside an alternative notification delivery schedule, users can opt in to an intentionally slower digital existence, primed for self-discovery and personal transformation. Siempo and intentional slowness stand in contrast to conventional digital consumptive behaviors such as habitual use, offering a new and *authentic* type of lifestyle or consumption practice as a means to resist the attention economy.

By providing an alternative practice of consumption, Siempo and intentional slowness reinforce disconnection as a type of neoliberal lifestyle politics (Kaun and Treré, 2018). Opting into intentional slowness places the impetus of change on the individual instead of industry and political elites to provide less invasive technology design practices. The user is encouraged to better manage their own time while the overriding imperative of UX design to tether users to their devices is more or less unquestioned. A future question for research could be whether intentional slowness provides a potential scapegoat for the largest poachers of our time and attention. In a June 2019 US Senate hearing on persuasive technology, Google UX lead Maggie Stanphill (2019) claimed that Google "supports an intentional relationship with technology," and in 2018 an Android feature called Digital Wellbeing was launched to enable users to set app restrictions to align with their intentions. Such intention-enabling features follow Siempo in disconnecting users from distractions but connect them to new regimes of consumptive practices that are positioned as key mechanisms of change. Perhaps to achieve significant reform of the so-called attention economy, scrutiny should shift away from the intentions of the user to the intentions of those who engineer user behavior.

NOTES

1. I choose "dropping out" instead of "opting out" because of the distinct ties to the historical Silicon Valley cultural practice of dropping out from mainstream society in the 1960–1970s and Timothy Leary's famous catchphrase.

2. Customization of the home screen is the primary functionality of Android launcher applications (Beal, n.d.). Launcher apps are android specific technologies that distinguishes android as an open software development platform to Apple's more tightly controlled app ecosystem. As Siempo is incompatible with Apple's app development policy, it is therefore only possible on android devices.

REFERENCES

Ash, James. 2015. *The Interface Envelope: Gaming, Technology, Power*. Oxford: Bloomsbury Academic.

Barbrook, Richard, and Andy Cameron. 1996. "The Californian ideology." *Science as Culture* 6 (1): 44–72. https://doi.org/10.1080/09505439609526455

Beal, Vangie. n.d. "Android Launcher." Webopedia. Accessed 1 May 2019. https://www.webopedia.com/TERM/A/android_launcher.html.

Beattie, Alex. 2018. "Out of Network." Real Life. May, 29, 2018. Accessed 28 August 2019. https://reallifemag.com/out-of-network/

Bosker, Bianca. 2016. "The Binge Breaker: Tristan Harris believes Silicon Valley is addicting us to our phones. He's determined to make it stop." The Atlantic. Accessed August 28, 2019. https://www.theatlantic.com/magazine/archive/2016/11/the-binge-breaker/501122/

Calvo, Rafael A., and Dorian Peters. 2014. *Positive Computing: Technology for Wellbeing and Human Potential.* Cambridge, MA: MIT Press.

Capsana. 2019. "Pause." Capsana. Accessed May 10, 2019. https://pausetonecran.com/en/manifest-home/

Carr, Nicholas. 2010. "Exodus." *Rough Type* (blog). May 8, 2019. http://www.roughtype.com/?p=1359

Cartwright, Darren. 2018. "'Digital detox' holidays are on the rise." Stuff. July 11, 2018, 2018. https://www.stuff.co.nz/travel/news/105391838/digital-detox-holidays-are-on-the-rise

Case, Amber. 2015. *Calm Technology: Principles and Patterns for Non-Intrusive Design*. Sebastopol, CA: O'Reilly Media.

CHT. 2019. "Design Guide (Alpha Version)." Center for Humane Technology. Accessed August 9, 2019. https://humanetech.com/designguide/

Craig, Geoffrey, and Wendy Parkins. 2006. *Slow Living*. Sydney: UNSW Press.

Cushman, Fiery. 2015. "Deconstructing intent to reconstruct morality." *Current Opinion in Psychology* 6: 97–103. https://doi.org/10.1016/j.copsyc.2015.06.003

Das, Salil K. 2003. "Harmful health effects of cigarette smoking." *Molecular and Cellular Biochemistry* 253 (1): 159–65. https://doi.org/10.1023/a:1026024829294

Dewe, Michaela, Jane Ogden, and Adrian Coyle. 2013. "The cigarette box as an advertising vehicle in the United Kingdom: A case for plain packaging." *Journal of Health Psychology* 20 (7): 954–62. https://doi.org/10.1177/1359105313504236

Dieter, Michael, and David Gauthier. 2019. "On the politics of chrono-design: Capture, time and the interface." *Theory, Culture & Society* 36 (2): 61–87. https://doi.org/10.1177/0263276418819053

Fish, Adam. 2017. "Technology retreats and the politics of social media." *triple C* 15 (1): 355–69.

Foucault, Michel. 1977. *Discipline and Punish: The Birth of the Prison*. London: Penguin Books.

———. 1997. "The ethics of the concern for self as a practice of freedom." In *Ethics: Subjectivity and Truth: Essential Works of Foucault, 1954–1984*, edited by Paul Rabinow, 281–302. New York: The New Press.

Ganesh, Maya Indira. 2018. "The center does not want your attention II. On time well spent and ethics." *Cyborgology* (blog), The Society Pages. February 15, 2018. https://thesocietypages.org/cyborgology/2018/02/15/the-center-does-not-want-your-attention-ii-on-time-well-spent/

Gauthier, François. 2013. "The enchantments of consumer capitalism: Beyond belief at the Burning Man Festival." In *Religion in Consumer Society: Brands, Consumers and Markets*, edited by François Martikainen Gauthier, Tuomas 143–58. Farnham: Ashgate.

Gilmore, Lee, and Mark Van Proyen, eds. 2005. *AfterBurn: Reflections on Burning Man*. Albuquerque: University of New Mexico Press.

Gregg, Melissa. 2018. *Counterproductive: Time Management in the Knowledge Economy*. Durham, NC: Duke University Press.

Guyard, Carina, and Anne Kaun. 2018. "Workfulness: Governing the disobedient brain." *Journal of Cultural Economy* 11 (6): 535–48. https://doi.org/10.1080/17530350.2018.1481877

Idhe, Don. 2008. "The designer's fallacy and technological imagination." In *Philosophy and Design: From Engineering to Architecture*, edited by P. E. Vermaas, P. Kroes, A. Light, and S. Moore, 51–59. Dordrecht: Springer Netherlands.

Kaun, Anne, and Emiliano Treré. 2018. "Repression, resistance and lifestyle: Charting (dis)connection and activism in times of accelerated capitalism." *Social Movement Studies*: 1–19. https://doi.org/10.1080/14742837.2018.1555752.

Kozinets, Robert, V. 2002. "Can consumers escape the market? Emancipatory illuminations from Burning Man." *Journal of Consumer Research* 29 (1): 20–38. https://doi.org/10.1086/339919

Lasch, Christopher. 1979. *Culture of Narcissism: American Life in an Age of Diminishing Expectation*. New York: Warner Books.

Leary, Timothy. 1999. *Turn On, Tune In, Drop Out*, 6th ed. Berkeley, CA: Ronin Publishing.

Lewis, Paul. 2017. "'Our minds can be hijacked': The tech insiders who fear a smartphone dystopia." *The Guardian*. Accessed April 9, 2019. https://www.theguardian.com/technology/2017/oct/05/smartphone-addiction-silicon-valley-dystopia

Light, Ben, Jean Burgess, and Stefanie Duguay. 2016. "The walkthrough method: An approach to the study of apps." *New Media & Society* 20 (3): 881–900. https://doi.org/10.1177/1461444816675438.

Malle, Bertram F., and Joshua Knobe. 1997. "The folk concept of intentionality." *Journal of Experimental Social Psychology* 33 (2): 101–21. https://doi.org/https://doi.org/10.1006/jesp.1996.1314

McGuigan, Jim. 2009. *Cool Capitalism*. New York: Pluto Press.

Odell, Jenny. 2019. *How to Do Nothing: Resisting the Attention Economy*. New York: Melville House.

Ofcom. August 4, 2016. Communications Market Report 2016. Accessed August 28, 2019. https://www.ofcom.org.uk/__data/assets/pdf_file/0024/26826/cmr_uk_2016.pdf

Ohtomo, Shoji. 2013. "Effects of habit on intentional and reactive motivations for unhealthy eating." *Appetite* 68: 69–75. https://doi.org/https://doi.org/10.1016/j.appet.2013.04.014

Orben, Amy, Tobias Dienlin, and Andrew K. Przybylski. 2019. "Social media's enduring effect on adolescent life satisfaction." *Proceedings of the National Academy of Sciences:* 201902058. https://doi.org/10.1073/pnas.1902058116

Paasonen, Susanna. 2016. "Fickle focus: Distraction, affect and the production of value in social media." *First Monday* 21 (10). https://doi.org/10.5210/fm.v21i10.6949

———. 2018. "Affect, data, manipulation and price in social media." *Distinktion: Journal of Social Theory* 19 (2): 214–29. https://doi.org/10.1080/1600910X.2018.1475289

Portwood-Stacer, Laura. 2012a. "How we talk about media refusal, part 3: Aesthetics." *Flow Journal*. Accessed May 1, 2019. https://www.flowjournal.org/2012/10/how-we-talk-about-media-refusal-part-3-aesthetics/

———. 2012b. "How we talk about social media refusal part 2: Asceticism." *Flow Journal*. Accessed April 27, 2019. http://www.flowjournal.org/2012/09/media-refusal-part-2-asceticism/

Rauch, Jennifer. 2018. *Slow Media: Why "Slow" Is Satisfying, Sustainable, and Smart*. New York: Oxford University Press

Reer, Felix, Wai Yen Tang, and Thorsten Quandt. 2019. "Psychosocial well-being and social media engagement: The mediating roles of social comparison orientation and fear of missing out." *New Media & Society* 21 (7): 1486–1505. https://doi.org/10.1177/1461444818823719

Royal Society for Public Health. 2018. "Scroll Free September." Royal Society for Public Health. Accessed May 10, 2019. https://www.rsph.org.uk/our-work/campaigns/scroll-free-september.html

Sharma, Sarah. 2014. *In the meantime: Temporality and Cultural Politics*. Durham, NC: Duke University Press.

Shenker, Barry. 2010. *Intentional Communities Ideology and Alienation in Communal Societies*. Boston: Routledge.

Siempo. 2019. "Welcome to Siempo." YouTube, April 9, 2019. Accessed December 15, 2019. https://www.youtube.com/watch?v=eNlj32aPasA&t

Skinner, B. F. 1948. *Walden Two*. Indianapolis, IN: Hackett Publishing Company.

Stanphill, Maggie. 2019. "Optimizing for engagement: Understanding the use of persuasive technology on internet platforms." US Senate Committee on Commerce, Science, and Transportation. Accessed August 28, 2019. https://www.commerce.senate.gov/public/index.cfm/2019/6/optimizing-for-engagement-understanding-the-use-of-persuasive-technology-on-internet-platforms

Steinhauer, Jillian. 2018. "The vanishing idealism of burning man." The New Republic, August 22, 2018. Accessed August 28, 2019. https://newrepublic.com/article/150497/vanishing-idealism-burning-man

Sutton, Theodora. 2017. "Disconnect to reconnect: The food/technology metaphor in digital detoxing." *First Monday*. https://doi.org/https://doi.org/10.5210/fm.v22i6.7561

Taplin, Jonathan. 2017. *Move Fast and Break Things: How Facebook, Google and Amazon Have Cornered Culture and What It Means for All of Us*. London: Pan Macmillan.

Turner, Fred. 2006. *From Counterculture to Cyberculture: Stewart Brand, the Whole Earth Network, and the Rise of Digital Utopianism*. London: The University of Chicago Press.

———. 2009. "Burning Man at Google: A cultural infrastructure for new media production." *New Media & Society* 11 (1–2): 73–94. https://doi.org/10.1177/1461444808099575

UPLIFT. 2015. "From Burning Man to Intentional Community." Accessed May 9, 2019. https://upliftconnect.com/burning-man-intentional-community/

Vaidhyanathan, Siva. 2018. *Anti-social Media: How Facebook Disconnects Us and Undermines Democracy*. New York: Oxford University Press.

Wajcman, Judy. 2019. "How Silicon Valley sets time." *New Media & Society* 21 (6): 1272–89. https://doi.org/10.1177/1461444818820073

Wakefield, Melanie A., Linda Hayes, Sarah Durkin, and Ron Borland. 2013. "Introduction effects of the Australian plain packaging policy on adult smokers: A cross-sectional study." *BMJ Open* 3 (7). https://doi.org/10.1136/bmjopen-2013-003175

Weiser, Mark, and John Seely Brown. 1995. "Designing calm technology." Accessed August 9, 2019. https://calmtech.com/papers/designing-calm-technology.html

Wu, Tim. 2016. *The Attention Merchants: The Epic Scramble to Get Inside Our Heads*. New York: Vintage.

Zandbergen, Dorien. 2012. "Fulfilling the Sacred Potential of Technology: New Edge Technophilia, Consumerism and Spirituality in Silicon Valley." In *Things: Material Religion and the Topography of Divine Space*, edited by Dick Meyer Houtman, Birgit, 356–79. New York: Fordham University Press.

Chapter Nine

"Life Hacking" Everyday Temporality

Project Managing Digital Lives of Tasks

Mikolaj Dymek

This chapter explores a central component of the so-called life hacking movement, namely task management and its material manifestations as mediatized technologies, which are predominantly digital applications/services that use project management approaches to target the liminal spaces of temporalities between work (professional/public/job life) and (domestic/private/spare time) life. The notion of "task" is nebulous within this movement—basically any future (planned) activity that can be written down and input into these systems is considered suitable for task management systems. The movement achieved recognition when in 2011 the Oxford Dictionary introduced the term "life hacking" as "a strategy or technique adopted in order to manage one's time and daily activities in a more efficient way" (Oxford Dictionaries, 2014). Since then a plethora of commercial task management apps and services have been released on desktops and smart devices reaching millions of life hacking mainstream users (e.g., popular task management app Trello in 2017 boasted about twenty-five million users; Pryor, 2019). The research question of this chapter is to explore the temporality discourses permeating this new generation of digital tools, and in particular in relation to relevant themes of liminalities in modernity, postmodernity, and hypermodernity. It will do so by focusing empirically on a particular case—OmniFocus 3, a system of personal task manager apps, and a pioneer in the commercial life hacking task management field.

BACKGROUND: LIFE HACKING AND TIME/TASK MANAGEMENT

Life hacking tools aim to increase efficiency of task execution and ultimately yield more spare time for its users, thus in some ways creating "spare time." It can be framed as targeting the work-life balance—a pivotal boundary of high modernity that separates remunerated (professional) work and (domestic) life. Work-life balance is a fairly well-established notion within human relations, business studies, organizational studies, just to mention a few. "Work" is the professional spheres of (predominantly) financially enumerated labor, whereas "life" is constituted by the "the rest," dominated by the private spheres of domestic and individual activities. By integrating all work *and* life tasks into task management tools, life quality will purportedly improve. This blurring of boundaries by amalgamation of "work" and "life" tasks/time, is also in equal measure dialectically maintained by the promise of a liminal space of increased spare time beyond and between work tasks and life errands.

Life hacker time/task management strategies can be nondigital, but (information) technology is often seen, within the community, as the preeminent enabler of efficiency—regardless whether digitalization is indeed necessary in terms of productivity (paper vs. digital personal calendars). This study will explore, in terms of (digitally) mediatized discourse dynamics, one of the most prominent digital tools for life hacking—OmniFocus 3.

Life hacking is a recent phenomenon with an unclear history. Admittedly, with (software and hardware) technological origins it approaches work and personal life as a "sub-optimized" technology system—components are analyzed in terms of efficiency and bottle-necks. British technology journalist Danny O'Brien (Trapani, 2005), observed so-called alpha geeks (self-proclaimed prominent computer enthusiasts/professionals) and how they "were able to process more information in a day than most literate 18th century readers would consume in a year" (Hogge, 2007). O'Brien's observations "went viral" on the internet turning quickly into a grassroots movement (Thompson, 2005) of life hackers.

Notwithstanding this heroic authorship view, the notion reached critical mass when combined with another time/task management movement—the slightly older *getting things done* (GTD) movement. GTD is a trademark of the David Allen Company whose namesake founder (Allen, 2002) introduced a technique that promised increased productivity and stress-reduction. The model stipulates five steps, where the smallest tasks take less than two minutes to complete—remaining tasks are organized according to various metadata. Two years later "indie writer" and inspirational speaker Merlin Mann (2007) launched the vibrant 43folders.com community that evangelised GTD and his own addendum—a filing system with thirty-one daily and twelve monthly folders (hence forty-three folders). This and other communities grew explosively, spawning numerous task management tools that noted technology magazine *Wired* claimed constituted a "new cult for the info age" (Andrews, 2005).

THEORETICAL FRAMEWORK

This chapter will contextualize a case of life hacker task management against a backdrop of relevant discourses within late/liquid modernity and hypermodernity. These discourses are centered around notions of time, work, media and power/organization (and in particular project management), and aspects that influence these (e.g., individualism, consumption, fashion, and flattened epistemologies). Furthermore, theories of liminality will be used to frame all these theoretical perspectives in order to further the understanding of the empirical material.

Task management à la life hackers is not geared towards organizational efficiency—mainly towards individual efficiency. As extensions of project management theories based on control, measurement and categorization of the world, they constitute a Habermasian (Habermas, 1987) *technological knowledge interest* focusing on desirable results by controlling variables in causal relations. Whereas these are attributes of the highest of modernities, there is a palpable context of liminal temporalities of liquid modernity (Bauman, 2000) to this phenomenon. With contemporary societies' public and private authorities in "liquid states" and (perceived) disorder, there are increased expectations towards "responsibility for failure falling primarily on the individual's shoulders" (Bauman, 2000, 8)—life hacker task management could be framed as an individualized response to manage these transformational challenges. Private solutions to public problems, promising not only answers in work, but *also* domestic life, yielding a liminal space of "saved time." Admittedly, this is a poignant case of the discursive time political effects of "disorganized capitalism" (Lash and Urry, 1994), reproducing the power—and time—relations of the capitalist industrial production systems.

However, these time relations seem to have been revitalized for the next generation of fragmented, "globish speaking" workforce of ever-flexible, mobile, teleworking, rootlessly ambulating freelancing individuals finding solace in the latest "app-ified" on-demand service amenities of the sharing "gig economy." This revised context requires an updated analysis. In these technological mediatized settings, the hedonistic instantaneous time frame of liquid modernity weariedly prevails, but in a significantly more anxious and emotionally conflicted fashion that begets interpretations by Lipovetsky's (2005) rewarding notion of *hypermodernity*. This notion affords several avenues of relevant temporal perspectives positing *consumerism* and *fashion* as the two dominant dimensions of contemporary society. These are consequences of the postmodern emancipation of the1960s where traditions, authority, and edifices of modernity were disassembled and rearranged into an age of ever-fluctuating individualism. It endures by means of insatiable consumption and it is recursively revitalised by the dynamics of fashion. Subsequently, temporal perspectives have shifted towards a specific instantaneity—the ephemerality of fashion and the hedonism of consumption venerates the present, forgos the past but is contemporaneously apprehensive of a dismal future plagued by the malfunctions of a global mass-consumerist society.

The notion of liminality is the consistent thread running through these theoretical approaches to life hacker task management. Theories of liminality (Coman, 2008; Czarniawska and Mazza, 2003; Küpers, 2011; Szakolczai, 2009; Waskul, 2005) based on the seminal works of van Gennep's (1960), Turner (1967) and Goffman (1959), will be used to frame this phenomenon throughout the analysis.

METHOD

Digital life hacker task management tools are engrained with discourses and meanings. While this is evident from a new/digital media studies perspective, it could be vigorously claimed that the analysis of these texts presents a formidable challenge for social sciences (Orlikowski, 2007). In many cases the analysis of digital media technology often transforms into discursive manifestations of existing theoretical fields. Digital media technology becomes a "container" of existing theoretical discourses, without extensively exploring the technological intricacies of *how* digital media, as such, affects and sometimes disrupts these discourses.

This study treats digital texts as empirical data in *itself.* Although appearing at first glance as empty grids of shiftable text containers, *meanings* are imbued inside these task management systems and their dynamic textual structures. The analysis shifts to the textual structures that users "*wread*"—*wr*ite and *read* as conceptualized by Landow (1994). Within new/digital/game media studies this methodological analysis evolved into polemics between viewing digital texts as (interactive) narratives (e.g., Laurel, 1993; Murray, 1997; Ryan, 2001) or as internal (medium) textual dynamics. A prominent theory of the latter perspective is Aarseth's (1997) *cybertext* (*cyborg text* machines) theory centered on notions of *ergodic texts* (dynamic texts that require mechanical input). Narrative dimensions are not excluded but it is considered more pertinent to analyse *what* happens on screens, *what* (interface elements) users modify, *how* (software) mechanisms modify text elements, and *how* hidden (not displayed) text elements are accessed (from databases). Aarseth basically provides a stringent deconstruction of the mythical notion of "interactivity" replacing it with the cybertext concept with very broad ramifications for digital media/texts analysis.

The cybertextual ergodic text perspective is limited by predisposition towards codex-like texts. Contemporary digital texts are seldom purely codex like and contain basically every mode of digitizable communication (*"multimedia"*). Consequently, methodological extensions that embrace multimodal communication is needed, such as *multimodal discourse analysis* (MDA) (Kress and van Leeuwen, 2001), where various semiotic modes are integrated in a discourse analysis adapted to contemporary media. Kress and Leeuwen criticise semiotics for fragmentation into incompatible "monomodal" semiotical perspectives, each telling their side of the story while missing the *entire* story. Kress and van Leeuwen (2001, 4) define four practice strata where dominant meanings are generated: *discourse* (socially constructed knowledge of reality), *design* (form conceptualizations of semiotic products and events), *production* (material form articulation of semiotic products or events) and *distribution* (reproduction of semiotic products and events). Implications of MDA stretch far beyond digital texts, but establishes the *design* stratum, in digital texts as the site of multimodal integration of semiotic modes. By methodological integrating cybertext and MDA perspectives, meanings can be discerned within the discourse strata as reflected in the design stratum based on the interplay between semiotic modes. Due to the duality of software (source/machine code) the strata of production and distribution are of lesser importance. A complete software analysis of material articulation requires source code access and is impractical in its scope (probably tens of thousands of code lines), and legally impossible (source code is proprietary). Consequently, a cybertextual MDA perspective incorporates textual mechanisms, *as well as* all other semiotic communication modes.

Cybertextual MDA methodology differs from alternative media new media approaches, such as walkthrough methodologies (Light, Burgess, and Duguay, 2018), as it prioritises the *dynamic* aspects of new media (traversal functions) and not merely "material traces of [designer] intentions" (Light, Burgess, and Duguay 2018, 886). It does, however, share its aim of critically examining the workings of an app as a sociotechnical artefact.

CASE STUDY: OMNIFOCUS 3

OmniFocus is a "task management platform" for computers, tablets, smartphones, and even smartwatches. This study performs a cybertextual MDA analysis of the most comprehensive desktop version. Similar to the walkthrough method's technical walkthrough procedure (Light, Burgess, and Duguay, 2018, 891) this analysis explores the app's interface, with focus on deeper, in terms of media technology, traversal functions (Aarseth 1997, 63).

The OmniFocus platform presents the user (see figure 9.1) with an expressive screen of text elements, graphics (symbol, icons, fields), various colors,

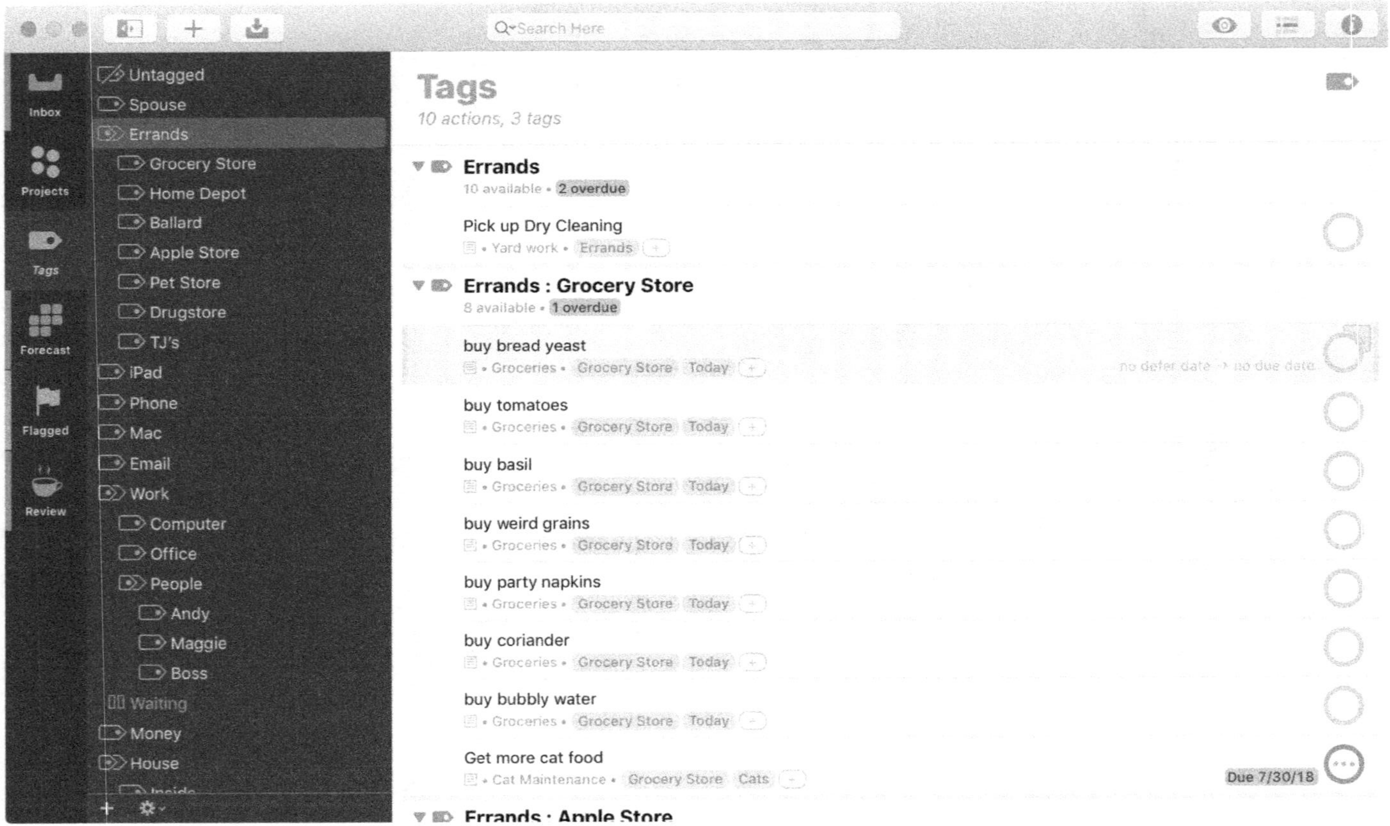

Figure 9.1. Interface OmniFocus

intricate graphical structures, movement/animation providing multimodal communication integrated in the *design stratum* (as per Kress and van Leeuwen, 2001) of this digital text. The core "interactive" text mechanism—*traversal function* using Aarseth's (1997, 63) notions—is investigated through a general walkthrough of the user workflow of OmniFocus 3.

The application is divided into most dominant traversal functions: "Inbox," "Projects," "Tags," "Forecasts," "Flagged," and "Review." "Inbox" constitutes the most fundamental traversal function for collecting text input (task descriptions), adding and categorizing related metadata. This task register is visually represented by a vertical list but also in horizontal relations (hierarchical categorization clusters of "Projects"). Every task can be complemented with metadata: "Title" (of the task as provided by user), "Status" (of task—active or completed), "Project" (hierarchial group of task), "Tags" (of task, user defined), "Estimated Duration" (of task), "Deferred Until" (postponement of task to later date by removing it from list of current task), and "Due" (date). These metadata are awarded separate traversal functions: "Projects," "Tags" and "Forecast." When task metadata is provided it is deferred (hidden) into the database structures. This constitutes the core task temporal organization functionality—to defer information *until needed again,* thus purportedly reducing information complexity and stress. In terms of multimodal communication, the vertical tasks list allows for organizing in visually horizontal relations—tasks clusters, "Projects." Lists can be "collapsed" and "expanded" with graphics animations—well-known since the invention of the graphical user interface. User expectations for hierarchical lists are shaped by previous graphical user interface conventions.

The "Projects" icon corresponds to related traversal functions allowing users to access/manipulate hidden tasks according to the metadata perspective of "Project." Similarly, as the inbox, list organization apply but are in terms of multimodal semiotical communication organized in hierarchical lists of *projects* (and folders/projects of projects). Project-specific metadata can be added here, such as "Repeat" (how often a project should be resurfaced), "Review" (date when to be reminded of a review of project which is explored with the "Review" traversal function) and "Note" (free text).

"Tags" provide traversal functions organized according to the metadata criterion "Tag." This traversal function allows for horizontal categorization of tasks according to user-defined tags that can also be organized in (visual/logical) hierarchies themselves. This user defined metadata represents physical and conceptual "spaces" where tasks exist, as defined by users. Typical tags are usually physical spaces ("Office" and "Home," see figure 9.1), temporal/conditional ("Waiting," "high energy"), persons ("Boss"), objects ("Car"), etc. Similarly, the "Flagged" traversal functions lists tasks that have been attributed a "flag"—basically a tag that signals importance, and is afforded a separate traversal function.

"Forecast" represents the most textually advanced traversal function with a mix of multimodal symbols—a visual calendar representation where each date square displays number of tasks with completion/or deferred dates on that day. These contain collapsible lists of tasks arranged in chronological order.

Finally, "Review" is a traversal function where, upon a system reminder, the user explores projects (hierarchically grouped tasks) in order to update project metadata, tasks, and to be reminded of overdue tasks/projects. This traversal function is similar to "Projects" as it organized in hierarchical lists of projects.

What this cybertextual MDA analysis of OmniFocus 3 has shown is that this case of task management apps is based on a user-created database whose access, usage, and interaction is delineated by the core traversal functions of "Inbox," "Projects," "Tags," "Forecasts," "Flagged," and "Review." These explore a database where the basic information and activity unit is a "task." This unit is apparently universal as it is supposed to suit any user, any setting, or work/life sphere. A group of tasks constitutes a "project," which can be attributed by any contexts (tags) that the user wishes, and most importantly, by due dates. The user is assumed to have no explicit social connections or communication within project activities beyond those than can be labeled as contextual tags. Tasks are hidden in the database and presented by the application to the user primarily when the due date is approaching. But tasks can also be explored according to user-created projects (hierarchically), tags/contexts (horizontally), calendar time (primarily future through forecasts), or when prompted by the system to review/manage projects.

ANALYSIS

The analysis of the OmniFocus 3 case will now turn to theories of liminality in (liquid/high) modernity and then hypermodernity in order to conclude with a framework of liminalities based on the *borderlands* notion (Fornäs, 2002).

Liminal "Me Time" Temporalities in Liquid Modernity

Liminality and transitional spaces have steadily increased in academic popularity (see Czarniawska and Mazza, 2003), or the special issue on transitional spaces by *Tamara Journal for Critical Organization Inquiry* (2011). Liminality has evolved from van Gennep's (1960) anthropological notions of *rites de passage* in small-scale societies, later developed by Turner (1967) and Goffman (1959) to include other types of transitional spaces.

In this study the liminality notion explores temporalities *between* the work *and* (domestic) life sphere. This temporality sphere is what life hacker task management tools aim to improve by reducing time spent on work *or* life tasks, thus yielding "saved time."

"Work-life balance" stems from the division of life into work and leisure, intrinsically linked to the industrial revolution (Guest, 2002) and modernity. In modernity, work and life constitute separate spheres of time, spaces, materiality, culture, organization, activities, dimensions, and values. The life sphere consists of the social contexts of the physical and symbolic notion of "home" and domestic dimensions—traditionally the context of the nuclear family.

Similarly, the work sphere has been radically transformed. Despite paramount transformations of both "home" and "life" spheres—they still persist at both ends of an ever-shifting scale. Their organization and balance is constantly in flux and depends on profession, country, culture, organization, legal system, and several other factors. This constitutes one of the most permanent boundaries of modernity. Labor unions, tax systems—even the quasi-coerced labor campaigns of communist regimes acknowledged the symbolical sacrifice of "life" in exchange for "state work" in its discourses. In terms of de jure conceptualizations, most democratic market economies subscribe to a clear-cut separation of both spheres.

A "market centric" view of work-life balance dominates many conceptualizations focusing on productivity regardless of context. Defining "spare time" as the "life" dimension of the work-life balance skew (e.g., perspectives on women's unpaid domestic labor). This is supported in political economy practice (e.g., the Northern European tradition of social democratic welfare states maintains a distinct separation of work and domestic life). The state aids and protects *the individual* from work sphere dysfunctions (e.g., unemployment insurance, labor laws, tax subsidized unions, etc.) *and* the domestic "life" sphere (e.g., various laws/mechanisms that protect the individual vis-à-vis traditional family functions, social security benefits, pension insurances, etc.). In many "traditional" societies liminal spaces exist between work and life, whereas in the secular, market liberal, democratic societies of the West have eschewed in large parts these liminal spaces, and reorganized contemporary life around a core of "paid work."

The few work-life transitional spaces (e.g., "third spaces" and various consumption spaces) have with time declined. The disappearance of "third spaces" has been thoroughly analyzed (Oldenburg, 1998; Putnam, 2001). New third spaces (e.g., coffeehouse chain Starbucks) created for lengthy leisure visits, have been transformed into workspaces by freelance workers. Other spaces have transformed into "after work" spaces—liminal in some regards but frequently related by location and social context to the work sphere. Various consumer spaces provide degrees of liminality, but as Bauman (2001) points out, consumption in liquid modernity is predominantly an individualistic enterprise—we shop, exercise in health clubs, and commute "alone in a group."

Globalization, privatization, individualization, information technology, social movements, and the onset of liquid modernity is transforming our lives in Bauman's view. Bauman targets a globally mobile "absentee landlord" style of capitalism, where corporate functions are outsourced to ephemeral entities whose modus vivendi is myopic profit maximization with a zealous cost-cutting imperative. To quote: "[C]apital has cut itself loose from its dependency on labour through the new freedom of movement undreamt of in the past. The reproduction and growth of capital, profits and dividends and the satisfaction of stockholders have all become largely independent from the duration of any particular local engagement with labour" (Bauman, 2000, 149). Where does this leave the (precarious) temporalities of local workers?

This dismantling of modernity has created an *interregnum* (Bauman's famous interpretation)—a power vacuum at the highest echelons spreading to most corners of society, and into individual worker temporalities. Life hacker task management tools become a "survival response"—a specific answer to a specific framing of individual hardships: the overflow of tasks in every sphere of lives but also the increasing synchronization, compression, and individualization of its time frames. Since the dawn of *oikonomia* through the giant Fordist factories of Taylorism and the evolution of the sprawling field of project management—task management has linked with authoritative responsibility and the rendering of hierarchical temporalities. In liquid modernity that link is decoupled. As Bauman (2000, 148) puts it: "Contemporary fears, anxieties and grievances are made to be suffered alone."

These aspects—blurring of work-life sphere, the decline of transitional spaces, increased pressure on the individual—are reflected in the OmniFocus 3 case, where tasks are universal units regardless of sphere, and the framing is extremely individualistic—the productivity of work or life tasks are to be managed alone.

TOWARDS A HYPERMODERN ANALYSIS OF TASK MANAGEMENT: NEW TEMPORAL LIMINALITIES

Digital media, however, is transforming our lives beyond the frames of liquid modernity, and possibly beyond chronotopia ("task management as time saver and productivity enhancer") vs. chronodystopia ("task management as a capitalist tool of oppression"). To move this inquiry to new relevant territories, attention is turned to the writings of French philosopher Gilles Lipovetsky and his stringent analysis of a so-called *hypermodernity* and its associated dynamics of temporalities. Of particular interest will be the dimensions of individualism, consumerism, instantaneous temporalities, and flattened epistemologies.

Individualism

> Individual autonomy appears increasingly as a norm imposed by organizations, but the life of choice, the do-it-yourself life, continues. The constraints of professional life are being reinforced, and the volatility of electors, couples, consumers and believers is becoming greater. (Lipovetsky and Charles, 2005, 80)

This reflects the empirical case of OmniFocus 3, which is exclusively personal, individual, and subjective by design and affordances. It does not exhibit any collaborative features or element. The user is an isolated individual in a world of unknown obstacles, vague tasks, and deadlines.

As discussed previously, it could favorably be analyzed as yet another example of capitalism's nefarious ways of imposing efficiency-centric temporalities with new ways of optimizing yet another sphere of postmodern labor. According to this view, users do not fully comprehend the reproduction of these power relations as they are beguiled by new task management tools often surrounded by hype, alluring promises of flashy life(style) transformations, and charismatic and well-meaning (start-up) creators. On global macro level this is undoubtedly an effective claim.

The rise of personal task management tools signals a decisive individualization of the project management field. From the mega-projects of high modernity, through the postmodern turn of applying project management to the (private) organization of the midsized postindustrial company in service industries, to possibly the last hypermodern step of applying the "project of project management" to an ever-smaller entity—is the individual's need to "remember the milk," which incidentally is also the name of a fifteen-year-old task management tool (Remember the Milk, 2019).

Lipovetsky outlines, however, that the individual passionately continues the "life of choice" as this is a direct continuation of the postmodern libertarian, anti-authoritarian, *and* anti-establishment emancipatory cultural revolutions of the 1960s (Lipovetsky, 2005, 38). Individualism is, it seems, worth the potential menace of oppressive norms.

Consumerism

Contemporary society is defined by consumerism, and OmniFocus 3 is no exception. Lipovetsky provides an explanation as to why:

> At the heart of the reordering of the way social time is organized lies the move from a capitalism of production to an economy of consumption, the replacement

> of an unbending and disciplinary society by a "society of fashion" restructured from top to bottom by the technologies of ephemerality, novelty and permanent seduction. (Lipovetsky, 2005, 36)

Lipovetsky's analysis (2005, 38) concludes that postmodern emancipation went astray, as meta-narratives were dismantled, and society found a new, yet highly precarious, shelter in the vagarious hands of consumerism and fashion—the two dominant dimensions of hypermodern society. The only thing that could sustain the hedonistic and individualistic drive of postmodern emancipation was consumerism and its ever-changing revitalization by logics of fashion. Hence, the postmodern individual challenge of authority "to find your own way" was most gratifyingly fulfilled by consumerism—new self-indulgent consumer products allowed new ways of self-expression and when those ways become well-trodden, a new fashion supplants previous product iterations with new waves. However, due to postmodernity's intrinsic unsustainability, in terms of (e.g., innovation, social atomization, communicative fragmentation, and not to mention environmental issues), this modus operandi was bound for an existential cul-de-sac. The hypermodern individual is partially aware of this complex unsustainability, causing a return of considerable weltschmerz in the postmodern pastures of hedonistic enjoyment. Nevertheless, the hypermodern individual is not able to escape the structures of his/her existence—consumption and fashion—and is therefore rummaging frenetically *inside* these structures for a solution.

Tools, such as OmniFocus, are preeminent examples of a consumerist approach to time, with elements of fashion logic included. The purported time-saving of task management tools is merely one app store purchase away (e.g., an OmniFocus 3 Pro license is $99.99 USD). For a movement targeting the ephemerality of time it is surprisingly material. Most systems are intended, although not required, to consist of hardware and software tools that are far beyond the financial reach of anyone outside the middle/upper socioeconomical spheres of the West. Various discourses of fashion intersect this phenomenon: hardware/software (software updates, two-year smartphone cycles etc.), management/organization ("lean," "agile," "big data" etc.), project management (Kanban, scrum, iterative development, etc.), digital culture ("social media detox," "screen time"), life-hacker subculture (GTD, time blocking, Pomodoro, etc.) are some of the fashion circuits affecting OmniFocus's field of task management. These discourses tirelessly revitalize the relevance of these tools, but also as components of consumer identity construction/expression processes.

Time is indeed money, as Benjamin Franklin famously defined this relation for modernity. In hypermodernity this logic has reached the everyday task management life of the individual. Lipovetsky points to a private hypermodern world invaded by corporate logic, *alongside* a self-reflective individualist "life of choice" with consumption, fashions, and self-aware weltschmerz. OmniFocus 3 promises to help deliver that pesky job project on time, but also reminds to buy that upcycled second-hand designer birthday gift for a friend. Time has become money, but money can passionately buy individualist enjoyment and self-expression through fashion.

Instantaneous Temporalities

> The loss of credibility of progressivist systems, the preeminence of the norms of efficiency, the commercialization of knowledge, the increasing number of temporary contracts in everyday life what could all of this mean if not that the centre of temporal gravity of our societies has shifted from the future to the present? (Lipovetsky, 2005, 35)

Much of the temporal analysis of postmodernity emphasizes the rise of instantaneous temporalities. Without grand narratives there is no relevant past, no history and the present is worshiped as a celebration of accelerating speeds in all aspects of life, compressed time frames complemented with fireworks of hedonistic pleasures. What changes in hypermodernity is, however, the nature of this instantaneity. The days of plentiful ideas of consumerism are over, and instead through the omnipresent consumerism and fashion logics, whose nihilistic openness towards anything novel, have opened up the troubled weltschmerz depths, pouring out to the boundaries of consumerism and its future—globalism, animal rights, climate change, wealth distribution, terrorism, eco-sustainability, populism, identity politics, public health, gender equality are just some of the issues affecting the future. The future is dark, uncertain, full of inconvenient truths that demand actions *now*—hence even further elevating the postmodern obsession of instantaneous temporalities of the present.

This is visible in OmniFocus 3. The past is deleted, archived and forgotten, whereas the future (through the traversal function of "Forecast") is a never-ending onslaught of stressing tasks—what is emphasized is how to deal with the present, and today's pressing tasks, through the other horizontally organized traversal functions ("Tags," "Projects," "Flagged"). OmniFocus 3 does not manifest the complex interconnections between past, present, and future tasks, as is the foundation of any traditional project management model, but instead the user is presented with various information mechanisms (traversal functions) of constructing "now"—the main objective of OmniFocus is to ideologically simplify complexity, stress, and uncertainty, thus allowing to focus on the urgent now. If postmodernity enjoyed the present with a positive outlook on the future, hypermodernity is even more focused on the present while having a tormented perspective on the future—task management tools are here to stress this situation.

Flattened Epistemologies

Task management is heavily indebted conceptually and as applied practice, to the notion of project management. While postmodernity questioned the grand narratives, in the age of hypermodernity the narratives are forgotten, and new ones are reinvented on novel and unexpected foundations:

> The dissolution of the unquestioned bases of knowledge, the primacy of pragmatism and the reign of money, the sense of the equal worth of all opinions and all cultures—these are all elements which feed into the idea that scepticism and the disappearance of higher ideals constitute a major characteristic of our epoch. (Lipovetsky, 2005, 67)

There is very limited continuity of more than one hundred years of project management theory in OmniFocus 3 despite having an explicit focus on projects. Fundamental basics of project management theory are omitted, such as the "project management triangle" where project quality is a trade-off between cost, scope, and time (Atkinson, 1999). Cost and quality dimensions are completely ignored, scope is somewhat acknowledged by the various categorization options of OmniFocus (tags and projects). The time dimensions, in terms of planning, is rudimentarily included—due dates, estimated duration, start dates are all optional, not necessary, planning details. These details form the basis of the Gantt chart, the ubiquitous project management tool from the turn of the twentieth century and onwards (Wilson 2003). This tool also prominently indicates dependency relationships between tasks—a prerequisite for any midsized (and upwards) project—something that OmniFocus, altogether overlooks.

By focusing on the present, forgetting the past and vexed by the future, OmniFocus 3 is seemingly more interested in providing an experience of stress relief, progress, and simplicity. Complexities, dependencies, time frames, social contexts, hierarchies, and competences are blissfully swept away in the name of "getting things done." Epistemologies of project management are flattened, or in some cases partially exchanged for surprising others (e.g., GTD's founder has expressed indebtedness to New Age cult-like Movement of Spiritual Inner Awareness). The task management movement doesn't care about project management theories, as long as the next trendy app achieves purported time savings.

Undoubtedly, flattened epistemologies are not something exclusive to the hypermodern universe, as the informal battle cry of Baudrillardian postmodernism (Braudrillard, 1983) postmodernism has been for decades *"there is only surface, only simulacrum."* Task management, in this view, is merely creating an *impression* of project management. Nevertheless, hypermodernism focuses on a particular flavor of this impression, namely the individual present and its need to deal with a vexing future timeframe.

LIFE HACKING THE WORK-LIFE BOUNDARIES AS LIMINAL TEMPORALITIES

Communication and cultural studies scholar Fornäs (2002, 93) proposes an analysis of liminal phenomena based on the *borderlands* notion from the works of Gloria Anzaldùa and James Clifford. Three borderlands types are distinguished: *free fields*, *battlefields*, and *crossing fields*. Applied to the work-life balance, free fields are isolated sanctuaries and free zones (e.g., the third spaces of yore usually confined to physical spaces (pubs, cafeterias, etc.)). As earlier analyzed they are disappearing as "spaces are liquefied"—mobile ICT transforms any physical space into a work/life space. A possible liminal candidate of free fields could be the popular concept of "me time" (Shaw, 2015)—to escape job and domestic chores and spend time alone (e.g., in "solitarity retreats"). It is, however, doubtful—despite what task management advertisements claim—that "me time" is enabled by productivity-enhancing task management apps. "Me time" could rather be understood as a manifestation of Lipovetsky's claims (Lipovetsky and Charles, 2005, 80) of supreme individualism in hypermodern times.

Battlefields are spaces of contradiction/fighting. Examples are political parties, labor unions, or (welfare) states operating in contested boundaries between work and life spheres negotiating boundary position (predominantly against monodirectional market economy forces, or political opponents). In some ways, OmniFocus 3 is a battlefield where work and life spheres are negotiated. Lipovetsky analyzes contemporary life as a dualistic existence of corporate logic *and* "life of choice." However, this negotiation has been the case for the worker since the dawn of the industrial age and is hardly something confined to the domain of OmniFocus 3. Furthermore, OmniFocus 3 blurs the boundary by unifying all spheres into one universal entity—the "task" position in the database, and by doing so enable Lipovetsky's dualistic descriptions.

Finally, crossing fields are those of transgressions, bridging opposites and hybridizations. OmniFocus 3 (aim to) enable these with productivity, it could be claimed. Transgressions are movements across borders (e.g., when in classical critical theoretical perspective "the bourgeois public" invades the "private spheres") (Habermas, 1991). Task management is viewed as a hegemonic tool of "corporate logic" enforcing "life efficiency." Liminal spaces sometimes appear as pseudo-transitions (Küpers, 2011) where false authorities, "tricksters" (Szakolczai, 2009) abuse influence. Perhaps task management is a trickster of capitalist project management intruding under the guise of stress-reduction (together with mindfulness and other self-improvement techniques)? The analysis of OmniFocus 3 indicated traversal functions of panoptical control—users facilitate OmniFocus control of task organization in work/life, promoting a "management of the self."

A crossing field that bridges opposites—a positive analysis of task management tools as enablers of borderlands of stress-reduced liminal spaces between work and life. By entrusting OmniFocus, users resolve tensions between work and life, gain control, reduce stress, yielding more spare time that can be used freely. Perhaps task management truly offers a bridge between work and life, thus generating new opportunities for some? Maybe complexity-reducing automation relieves users from the stress-inducing spheres of "work" and "life"? Admittedly, OmniFocus reduces information complexity and most likely reduces *some* stress, but is it fair to characterize this as completely bridging opposites? However, if assuming that task management tools provide perceived or quantitively measurable time savings, this is countered by the so-called Jevons' Paradox, outlined by nineteenth-century economist William Stanley Jevons (Alcott, 2005), that states that an increase in resource efficiency leads to increased resource consumption rather than reduction. Tasks and time are hardly "resources," nevertheless the logic still applies—if faced with increased "spare time" away from work and (domestic) life due to task management, will users: (a) enjoy the stress-free liminality or (b) undertake *more* tasks? In contemporary society, to make a long answer short, the option is seemingly b.

Finally, hybridization enables the bricoleur (Lévi-Strauss, 1966) of liquid modernity to overlap what is elsewhere separated. By amalgamating *all* tasks in one management tool the spheres of work and life are integrated. The empirical analysis showed that the "Tags" traversal function manipulates the name of the sphere with one click—providing a liquefaction of spheres. OmniFocus reduces and flattens life challenges into (arbitrary) units of tasks, task clusters/projects, deadlines/durations, and contexts/locations. Inevitably this leads to blurring of boundaries, hybridization, and new temporalities. Life hackers of the "Californian ideology" (Reagle, 2019, 149) (i.e., Silicon Valley professionals), certainly seem to believe so. This is also supported by Lipovetsky's flattened epistemologies argument—maybe hypermodern life makes no distinction between work and life, and OmniFocus embraces aspects of this new hybridization.

CONCLUSIONS

As a departure point, this study analysed OmniFocus 3 by means of cybertextual MDA in a discourse context of liquid modernity, hypermodernity, and borderlands of liminal temporalities with categories of "crossing fields" of transgressions, bridging opposites, or hybridizations. Basically, the question at hand is whether OmniFocus 3 offers liminal temporalities that, in essence, are enforcing capitalist "life efficiency" temporalities (transgressions), or offering improved stress-reduced temporalities (bridging opposites through complexity-"decluttering" task management), or blurring of work/life time boundaries to hybrid postmodern freelance-lifestyle temporalities? One might argue that the hybridization borderland category most rewardingly describes the aims of the life hacker community, but that casual users might subscribe to the bridging opposites category of liminal temporalities. Life hacking entails that life *is* hackable, whereas casual users wouldn't mind some stress-reduction in their lives and leave the hacking to others.

Lipovetsky's (2005) notion of hypermodernity—a postmodernism in hyperdrive, but also in retreat—applied on OmniFocus 3 provided perspectives on individualism, consumerism, instantaneous temporalities, and flattened epistemologies. Users are precariously isolated when faced with the complexity of everyday life, including corporate logics—but individualism is also an intrinsic part of the "life of choice" guided by consumption and fashion. Likewise, this duality applies to the remaining hypermodern dimensions discussed. Task management is part of the consumerist and fashion cycles—but this also because the hypermodern citizen uses these for self-expression and enjoyment. Hypermodern wo/man uses task management tools to focus even more on the present, because the future is vexing, partially because of global threats, but also because the hedonistic careless future of postmodernism has been fundamentally revised with substantial dosage of existential weltschmerz. And finally, the hypermodern consumer doesn't care/know whether the task management tools are based on scientific inquiry or New Age mysticism—as long as they provide a feeling and impression of progression in managing tasks and projects.

Basically, the conclusions here perhaps paint a picture of OmniFocus, and task management, beyond the dichotomous perspectives of chronotopia ("task management as time saver and productivity enhancer") vs. chronodystopia ("task management as a capitalist tool of oppression") in relation to time. The picture requires more nuances and this chapter's analysis has contributed towards that contextualization.

REFERENCES

Aarseth, Espen J. 1997. *Cybertext—Perspectives on Ergodic Literature*. Baltimore, MD: Johns Hopkins University Press.

Alcott, Blake. 2005. "Jevons' Paradox." *Ecological Economics* 54 (1): 9–21.

Allen, David. 2002. *Getting Things Done: The Art of Stress-Free Productivity*. New York: Penguin.

Andrews, Robert. 2005. "GTD: A New Cult for the Info Age." Wired. http://archive.wired.com/culture/lifestyle/news/2005/07/68103?currentPage=1.

Atkinson, Roger. 1999. "Project Management: Cost, Time and Quality, Two Best Guesses and a Phenomenon, Its Time to Accept Other Success Criteria." *International Journal of Project Management* 17 (6): 337–42. https://doi.org/10.1016/S0263-7863(98)00069-6.

Baudrillard, Jean. 1983. *Simulations*. New York: Semiotext.

Bauman, Zygmunt. 2000. *Liquid Modernity*. Cambridge: Polity Press.

———. 2001. *The Individualized Society*. Cambridge: Polity Press.

Coman, Mihai. 2008. "Liminality in Media Studies: From Everyday Life to Media Events. Victor Turner and Contemporary Cultural Performance." In *Victor Turner and Contemporary Cultural Performance*, edited by Graham St. John, 94–108. New York: Berghahn Books.

Czarniawska, Barbara, and Carmelo Mazza. 2003. "Consulting as a Liminal Space." *Human Relations* 56 (3): 267–90.

Fornäs, Johan. 2002. "Passages across Thresholds: Into the Borderlands of Mediation." *Convergence* 8: 89–106.

Gennep, Arnold van. 1960. *The Rites of Passage*. Chicago: University of Chicago Press.

Goffman, Erving. 1959. *The Presentation of Self in Everyday Life*. New York: Random House.

Guest, David E. 2002. "Perspectives on the Study of Work-Life Balance." *Social Science Information* 41 (2): 255–79. https://doi.org/10.1177/0539018402041002005.

Habermas, Jürgen. 1987. *Knowledge and Human Interests*. Cambridge: Polity Press.

———. 1991. *The Structural Transformation of the Public Sphere: An Inquiry Into a Category of Bourgeois Society*. Cambridge, MA: MIT Press.

Hogge, Becky. 2007. "Resolution One: Get a Life Hack; For Alpha Geeks, Every Day Offers a Chance to Improve Themselves." *New Statesman* 136 (4826).

Kress, Gunther R., and Theo van Leeuwen. 2001. *Multimodal Discourse: The Modes and Media of Contemporary Communication*. London: Hodder Arnold.

Küpers, Wendelin. 2011. "Dancing on the Līmen: Embodied and Creative Inter-Places as Thresholds of Be (Com)ing: Phenomenological Perspectives on Liminality and Transitional Spaces in Organisation and Leadership." *Tamara Journal for Critical Organization Inquiry* 9 (3–4). http://tamarajournal.com/index.php/tamara/article/view/139.

Landow, George P. 1994. *Hyper/Text/Theory*. Baltimore, MD: Johns Hopkins University Press.

Lash, Scott, and John Urry. 1994. *Economies of Signs and Space*. Thousand Oaks, CA: Sage.

Laurel, Brenda. 1993. *Computers as Theatre*. Boston: Addison-Wesley.

Lévi-Strauss, Claude. 1966. *The Savage Mind*. Chicago: University of Chicago Press.

Light, Ben, Jean Burgess, and Stefanie Duguay. 2018. "The Walkthrough Method: An Approach to the Study of Apps." *New Media & Society* 20 (3): 881–900. https://doi.org/10.1177/1461444816675438.

Lipovetsky, Gilles. 2005. "Time Against Time: Or The Hypermodern Society." In *Hypermodern Times*, edited by Gilles Lipovetsky and Sébastien Charles. Cambridge: Polity Press.

Lipovetsky, Gilles, and Sébastien Charles. 2005. "Stages in an Intellectual Itinerary: A Conversation between Gilles Lipovetsky and Sébastien Charles." In *Hypermodern Times*, edited by Gilles Lipovetsky and Sébastien Charles, 153–71. Cambridge: Polity Press.

Mann, Merlin. 2007. *About: 43 Folders*. http://www.43folders.com/.

Murray, Janet. 1997. *Hamlet on the Holodeck: The Future of Narrative in Cyberspace*. Cambridge, MA: MIT Press.

Oldenburg, Ray. 1998. *The Great Good Place: Cafés, Coffee Shops, Bookstores, Bars, Hair Salons, and Other Hangouts at the Heart of a Community*. New York: Marlowe.

Orlikowski, W. J. 2007. "Sociomaterial Practices: Exploring Technology at Work." *Organization Studies* 28 (9): 1435–48. https://doi.org/10.1177/0170840607081138.

Oxford Dictionaries. 2014. *Definition: Life Hack*. http://www.oxforddictionaries.com/definition/english/lifehack.

Pryor, Michael. 2019. "25 Million! Celebrate with 4 New Top-Requested Trello Features." *Blog.Trello.Com* (blog). December 9, 2019. https://blog.trello.com/25-million-users.

Putnam, Robert D. 2001. *Bowling Alone: The Collapse and Revival of American Community*. New York: Simon and Schuster.

Reagle, Joseph M. Jr. 2019. *Hacking Life*. Cambridge, MA: MIT Press.

Remember the Milk. 2019. "About the People behind the Craziness." Remember The Milk. 2019. https://www.rememberthemilk.com/about/.

Ryan, Marie-Laure. 2001. *Narrative as Virtual Reality: Immersion and Interactivity in Literature and Electronic Media. Parallax (Baltimore)*. Baltimore, MD: Johns Hopkins University Press.

Shaw, Gina. 2015. "Finding Time for 'Me' Time." WebMD. September 4, 2015. https://www.webmd.com/women/guide/womans-guide-to-me-time.

Szakolczai, A. 2009. "Liminality and Experience: Structuring Transitory Situations and Transformative Events." *International Political Anthropology* 2 (1): 141–72.

Tamara Journal for Critical Organization Inquiry. 2011. "Special Issue on Organizing Transitional Space & Special Issue on Aesthetics and Ethics (Double Issue)." *Tamara Journal for Critical Organization Inquiry* 9 (3–4).

Thompson, Clive. 2005. "Meet the Life Hackers." *New York Times*. https://www.nytimes.com/2005/10/16/magazine/meet-the-life-hackers.html.

Trapani, Gina. 2005. “Interview: Father of ‘Life Hacks’ Danny O’Brien.” Lifehacker. 2005. http://lifehacker.com/036370/interview-father-of-life-hacks-danny-obrien.

Turner, Victor Witter. 1967. *The Forest of Symbols: Aspects of Ndembu Ritual.* Ithaca, NY: Cornell University Press.

Waskul, Dennis D. 2005. “Ekstasis and the Internet: Liminality and Computer-Mediated Communication.” *New Media & Society* 7 (1): 47–63. https://doi.org/10.1177/1461444805049144.

Wilson, James M. 2003. “Gantt Charts: A Centenary Appreciation.” *European Journal of Operational Research, Sequencing and Scheduling*, 149 (2): 430–37. https://doi.org/10.1016/S0377-2217(02)00769-5.

Chapter Ten

Managing the Flow of Time

Disconnection through Apps

Carla Ganito and Cátia Ferreira

Digital technologies have a profound impact on how connected users perceive time. One could argue that we are witnessing a remediation of time as an objective and tangible element of everyday life. With the assertion of digital media, time is being shaped by new social and communication practices, contributing to a renegotiation of its meaning.

Manuel Castells (1996) calls this emerging temporality "timeless time," referring to the breach of sequentiality in social actions, be it by the compression of time, or by the random reordering of moments. The ordering of moments has lost its chronological rhythm and is now organized in temporal sequences conditioned by its social use or purpose.

In a "postdigital age" (Cramer, 2015; Thorén et al., 2017) where internet access is widespread, ubiquitous, and permanent, and where technologies are being integrated into objects, through IoT (Internet of Things) initiatives, digital media, and particularly social media, may be experienced as more invasive. They elicit users to use more time through constant notifications, and thus we witness the surfacing of practices of resistance and the revival of "old" media: "The post-digital era is, in comparison with the digital era, less linear, more cyclical; less efficient, more nostalgic" (Thorén et al., 2019, 328). This emerging resistance is also performed against a neoliberal discourse that places disconnection as a threat (Chandler and Reid, 2016).

An emergent field of research looks at the nonuse of digital media (Hesselberth, 2018; Kaun et al., 2016), particularly at people avoiding and quitting specific social media platforms, content and technologies (Light and Cassidy, 2014; Portwood-Stacer, 2012), actively choosing not to participate online (Kaun and Schwarzenegger, 2014), seeking "digital detox" and abstaining from using digital media in search of human presence (Karlsen and Syvertsen, 2016; Woodstock, 2014).

There are several type of measures to decrease or control internet use, from behavioural measures such as setting time limits, to technically based controls (Ganito and Jorge, 2018; Ofcom, 2017). In this chapter we look at the discourses of apps to disconnect or manage the use of online media (such as Freedom, Anti-social, Rescue Time, Disconnect, Deseat. me, Checky) and screen time functions from operating systems as technical tools to which users can resort to manage their on- and offline time (Guyard and Kaun, 2018). We seek to understand their different possibilities and how they construct and value time spent offline and, conversely, online. Specifically, we analyze their functioning design, as well as their claims and publicity discourses on news media. Their claim ranges from the promise of disconnection to improve productivity, to a promise of helping to counterbalance an always on, addictive digital time culture. Connectivity being a central element of "social media logic" (Van Dijck and Poell, 2013), applications and social media resort to gamification strategies to keep users engaged. Gamification is a concept that is related to the use of strategies of games in contexts and environments not directly related to the games (Deterding et al., 2011, 2). From an applied point of view, gamification strategies make use of elements traditionally found only in games, like progress bars, badges, points, or other reward systems, levels, challenges, boards, notification systems, etc. Most of these features configure an attempt to increase influence and engagement, as well as to make processes of exit or interruption more difficult to perform. The difficulty tends to be centered on a user's involvement and active engagement, which is made more palpable through the use of game elements and mechanics, since they have an emminent role in transforming the user of these platforms in a more active participant, an element that is crucial to the digital experience. All these aspects tend to be pervasive in digital platforms, and have been key in the design of websites, social media, and apps, with the purpose of engaging users and "keeping them in the loop."

We critically discuss these technical services as tools for increased productivity and controlled management of time spent online, but also as part of a governmentality discourse where the individual user is positioned as responsible for searching for and using services to self-regulate his/her use of digital media.

NEGOTIATING TIME

Today, it is expected of most of us to be available at any time and at any place. This is captured in what Castells (1996) has called "timeless time."

Timeless time translates into instantaneity or the speeding up of events with desequencing encapsulating a "perpetual present," which may in turn be grasped in the breaking down of "rhythmicity" through the technological transformations of the life cycle. He argues,

> timeless time, as I have labelled the dominant temporality of our society, occurs when the characteristics of a given context, namely, the informational paradigm and the network society, induce systemic perturbation in the sequential order of phenomena performed in that context. This perturbation may take the form of compressing the occurrence of phenomena, aiming at instantaneity, or else by introducing random discontinuity in the sequence. (Castells, 1996, 464)

The mere fact that one has the possibility to communicate with what one can call "the absent others" (Giddens, 1990), and thus has the possibility to go beyond the need to coordinate actions beforehand, gives technologies, especially mobile technologies, a strong role in the restructuring of time. It is what Anthony Giddens (1990) calls the "disembedding" of social interaction that is the central issue in this matter. A fundamental change in the notion of time happens when it is possible to exist in a communication-sphere regardless of spatial boundaries. The coordinating aspects of clock-time are put under pressure from the ever present and dynamical restructuring and renegotiation aspects of digital technologies. The linear time—as in clock-time—is not changed, what is changed is how the linear time is filled with actions (Johnsen, 2001, 63). We are available wherever we are and we can even resuscitate time, as the one we spend in transportation or in waiting lines. The practice of using the mobile phone to build a personal cocoon enables people to "transform 'dead time' in incidental locations into time that is personally productive or enriching" (Ito et al., 2008, 74). Digital technologies also allow for permanent multitasking. Manuel Castells (2008) designates multitasking as the "blurring of time and space" and shows how the mobile phone has become the tool of choice for multitasking by enabling the hybridization of spaces and the blurring of sequences in time. Southerton and Tomlinson (Southerton, 2006; Southerton and Tomlinson, 2005) designate this pattern of allocation of tasks as "temporal density," which together with "temporal dis-organization" and volume constitutes the basis for the sense of always running out of time. The acceleration of time is also pointed out as a paradox in face of the acceleration of technology (Rosa, 2003). At the same time, we must acknowledge that "ours is not the first era where overload is raised as an issue" (Syvertsen and Enli, 2019, 7) as with every media technology from books to television came the concerns and media panics over loss of control. Digital detox and time management tools feed on this discourse.

The role technology plays in this process of time poverty and in the perception that time is saved or lost has been very controversial. Han (2015) argues that time is not accelerated but dispersed:

> Today's temporal crisis is caused by a dysnchronicity which leads to various temporal disturbances and irritations. Time is lacking a rhythm that would provide order, and thus it falls out of step. The feeling that life is accelerating is really the experience of a time that is whizzing without a direction. (7)

Han also traces the main cause of today's temporal crisis to the value placed on a productive life, the "vita active": "the hyperkinesia of everyday life deprives human existence of all contemplative elements and of any capacity for lingering. So-called strategies of deceleration do not overcome this temporal crisis; they even cover up the actual problem. What is necessary is a revitalization of the *vita contemplative*" (Han, 2015, 7). Contrary to Arendt (1958), Han considers that "human action is reduced to mere activity and labour precisely by losing all of its contemplative aspects" (100). From a co-constructivist perspective, it is easy to understand that their impact would never be straightforward because "technologies change the nature and meaning of tasks and work activities, as well as creating new material and cultural practices" (Wajcman, 2008, 66). Acknowledging the agency of users allows us to understand innovative uses of technologies such as the applications that we carry in our mobile devices, to take control of time.

With the mobile phone, especially with web-enabled smartphones, we are in "perpetual contact" (Katz and Aakhus, 2002), in a state of "constant availability" (Chayko, 2008). This constitutes a paradox because, if on the one hand that helps to release the anxiety and to feel more connected to our network and thus emotionally rewarded, on the other hand it leverages our multitasking and burdens our daily life with tasks from different spheres, personal and professional.

The widespread possibility of constant connectivity has generated a social expectation that everyone should be available and when people choose to be out of touch or end up being out of touch by accident, it triggers criticism and sometime even self-criticism or remorse for having chosen to be out of reach (Ganito, 2016). This perception leads users to engage in creative forms of avoidance, and it also justifies the popularity of time management apps that proliferate in our desktops but mainly in our mobile devices.

METHODS

Although studies show that people tend to rely more on decision-based than technical or consumption measures to decrease their internet use (Ganito and

Jorge, 2018; Ofcom, 2016), there is an array of apps offering users "help" to disconnect or to manage the use of online media.

In order to better understand the offer related with connect/disconnect balance, attention will be paid to the discourses of web and mobile apps that promote a better management of the time spent online and/or users' online presence, aiming at understanding how they construct and value time spent offline. The primary data collection method was content analysis, which was complemented with firsthand experience of the apps and document analysis. The research variables were four: Discourses—how the app is presented; Functioning design (interface and available features); Usage (downloads, rating); and Media coverage of this type of apps.

The sample consists of ten web and mobile apps, organized in three categories: productivity (four), privacy (four), and unplug (two). These categories were defined after a preliminary analysis of the description of the apps. The privacy category is the less self-descriptive one taking into consideration that the empirical research is centered in time management apps. In this scope privacy category reflects the idea of resisting to the datafication of time, reinforcing the right to be forgotten (Couldry and Hepp, 2016; Couldry and Mejias, 2019). These apps target users that seek the possibility of controling the visibility and availability of data concerning the time they spend online. Specifically, we have analysed: Freedom, Rescue Time, Cold Turkey Blocker, FocusMe, Disconnect, Deseat.me, DuckDuckGo, SafeShepherd, Checky, and Binky. The first four apps are focused on users' productivity: Freedom and Rescue Time are originally web browser application that now are also available for mobile devices (android and iOS); Cold Turkey Blocker is a webapp; and FocusMe besides being available as a web browser application has also an android app. In the privacy category the analyzed apps are: Disconnect offers different products for different devices (browsers, iOS and android); Deseat.me and SafeShepherd are webapps; and DuckDuckGo being a search engine is specifically a web database application. The decision to include this search engine in the sample was because it offers users the opportunity to perform one of the most popular online activities (Statista, 2019)—search—in an application that does not colonize users' data and in extention the users' time, contributing to users' right to be forgotten. The two products analyzed in the third category, unplug, are: Checky and Binky mobile devices apps available for android and iOS. The choice of this sample was both purposive, which is aligned with qualitative research, and comprehensive as it reflects the most popular apps for monitoring and/or changing web and mobile media practices at the moment of data collection (September-October 2019). The corpus was selected taking into consideration two main aspects: the apps position in the app stores' rankings and opinion articles in technology publications. The search criteria was the phrase time management apps. In what the rankings are concerned, for instance, Freedom counts with more than one million users of the web app, according to the information available on the website and with more than fifty thousand registered installs in play store; Disconnect counts with more than one million users, according to the information on the app's website; Rescue Time has more than one hundred thousand installs in play store; and Checky has been downloaded from play store more than one hundred thousand times. We will start by presenting the claims that address three main concerns: productivity, privacy, and disconnection. We will then focus on the interface, and finally on the media discourse and how they present these apps.

RESULTS

Time management apps have built a discourse of reclaiming control over your time using three main claims: productivity, privacy, and disconnection. The four apps of the productivity category offer users a way to stop distractions in order for them to gain productivity. All of them present distractions as something that needs to be controled in order to liberate users. Distractions are presented as being the problem and productivity, focus, and balance what will be gained: "Stop being distracted by your laptop Freedom is the app and website blocker used by over 750,000 people to reclaim focus and productivity. Experience the freedom to do what matters most" (Freedom); "Meet your match, Zuckerberg. Boost your productivity and reclaim your free time by blocking distracting websites, games and applications" (Cold Turkey Blocker); "take back control of your day" (RescueTime); "Stop wasting your time on distractions" (FocusMe).

Concerning the second category, privacy, the discourses are centered on the idea of online tracking as something that puts users' privacy at risk. Web tracking and personal information that is crumbled over the web are the identified problems, which once solved would give users the opportunity to improve their privacy protection: "Disconnect defends the digital you. Say no to mass collection of your online activity and trackers that destroy your device performance" (Disconnect); "Clean up your online presence" (Deseat.me); "Protect yourself from the Internet. Safe Shepherd pro-actively removes your personal information from the Internet and marketing databases" (Safe Shepherd); "The search engine that does not track you" (DuckDuckGo).

The discourses of the apps of the third, and last category, unplug, are similar in what concerns the identification of the problem—distractions, but different in what they promise, since the two analyzed apps are much different from the rest of the sample. These apps offer help to counterbalance an always on, addictive culture. Checky starts by asking "How many times do you check your phone?" It is a counter; it counts how many times one checks the mobile phone; and Binky looks exactly as a social media platform but there are no friends or networks. Instead, content is produced by the Binky team and users may freely scroll their feeds and interact with content, but there is no network of contacts or interests, since it is not a networked app in that sense. Each user interacts primarily with content, not with other users. Both applications promise the opportunity to set a different relationship with one's mobile phone and Checky additionally promotes the increase of self-awareness regarding the mobile phone usage.

The interface is the main mediation element connecting users and cyberspace. The strategies used in the design of interfaces are very important, since they determine user experience, thus the second variable to be discussed will be functioning design, which implied analyzing apps' interfaces and features. A main conclusion is that the apps follow the same logic they try to counterbalance, one of simplicity and engagement (see Table 10.1).

Freedom is a web and app blocker with a simple and intuitive interface. The navigation and interaction with the menus is easy and the app configuration does not require much time. Users need to choose what they want to block, when, and in which devices. The web browser and mobile applications allow one to manage all devices where Freedom is installed. The main features are the customizable blocklists, the synchronization of blocks across devices, and the schedule tool to manage when Freedom should be active. RescueTime is a productivity report tool and website blocker also with a user-friendly interface. The navigation is intuitive in the apps for different devices; the use of visual elements helps to read the reports quickly; the configuration process is simple. It is available as a basic and a premium version. The free version includes the following basic features: detailed reports, weekly email summary, goals for the day, and productivity scores. Cold Turkey Blocker has a simple and intuitive interface. The websites blocker configuration is done in steps and it is easy to do. It is available as a free light version and a paid premium one. The basic features are: unlimited website blocking, unlimited block duration, motivational block page, locked mode, compatible with VPNs, statistic reports. FocusMe is a website and application blocker and a productivity report tool. The interface is simple, despite the number of elements that can be customizable. The use of colour is considered to be important since it makes reading all the information easier, particularly in what concerns the management of the "focus plans" (terminology used in the application) since all blocked websites and applications appear in red and all non-blocked in green. The available features are organized in two sets—productivity and blocking. For productivity: scheduler, break reminders, timer. For blocking: block websites and apps; block rules and a white list customization. All four productivity apps analyzed offer similar features according to the service offered—blocking or activity tracking. And all offer simple and user-friendly interfaces, despite offering users the opportunity to customize the application.

Table 10.1. Functioning Design Variable: Main Results

App	*Features*	*Category*	*Interface*
Cold Turkey Blocker	Basic features: unlimited website blocking, unlimited block duration, motivational block page, locked mode, cross-browser support, compatible with VPNs, start blocks from system tray, statistics	Productivity	Simple and intuitive interface; the websites blocker configuration is done in steps and it is easy to do
FocusMe	Available as web app for computers and android app for mobile devices; For Your Productivity: scheduler, break reminders, pomodoro timer most powerful blocking: block websites and applications, wild card allow/block rules and white list, blocking is instant, impossible to bypass; Time Tracking	Productivity	The interface is simple, despite amount of elements that can be customizable; the use of color is important since it makes reading all the information easier
Freedom	Compatible with Mac, Windows, iPhone, iPad; browser extensions (Chrome, Firefox, Opera); it blocks websites, apps, internet, customizable blocklists, sync blocks across devices, schedule; it is available in a mobile app version for iOS and android	Productivity	Simple and intuitive; the navigation and interaction with the menus is easy and the app configuration does not require much time, users need to choose what they want to block, when, and in each devices; the web and mobile apps allow one to manage all devices where Freedom is installed
Rescue Time	Tracks time spent on applications and websites giving you an accurate picture of your day. Basic features: detailed Reports, weekly email summary, set goals for the day, productivity score	Productivity	User-friendly; navigation is intuitive, the use of visual elements helps to read the reports quickly; the configuration process is simple

Deseat.me	Identification of registered accounts, management of the accounts (which to delete and which to keep), send removal requests	Privacy	Really simple, decision-making is presented with visual elements which makes the interface very intuitive
Disconnect	Block 2,000+ tracking sites, load pages 27% faster; compatible with Chrome, Firefox, Safari, Opera (browsers); features: requests management by topic; visualize page; white lists, show counter, cap counter	Privacy	Easy and simple, there is not much to customize, only a white list of blocking exceptions; the interface is intuitive
DuckDuckGo	Search engine that blocks ads, private search activity, control of your personal data	Privacy	Interface simple and really user-friendly; the use of color makes the information interpretation intuitive; the use of graphic elements helps navigating through the options
SafeShepherd	Privacy monitoring, instant privacy alerts	Privacy	Simple interface; there are several customization possibilities, the navigation and configuration is intuitive, but there is a lot of information to read
Binky	Social media look and features, but no one has access—there is no network of contacts to follow or that will be following you	Unplug	Visually appealing and its interface is really simple since there are not many customization options
Checky	Mobile devices app—android, iOS; category: Productivity	Unplug	Simple and visual interface; there is not much to customize or to explore, since it is a really simple app

As for the privacy category apps: Disconnect is a browser app that tracks and blocks websites that track users. It is easy and simple, there is not much to customize, only a white list of blocking exceptions; the interface is intuitive. The application's main features are: requests management by topic; visualize page; white lists; show and cap counters. Deseat.me is a digital accounts' cleaning tool, which is simple to use. Decision-making is presented with visual elements that makes the interface very intuitive. Its features are: identification of registered accounts, management of the accounts (which to delete and which to keep), and send removal requests tool. DuckDuckGo is a search engine that blocks advertisements and it has a simple and user-friendly interface. The use of color makes the information interpretation intuitive, since users' privacy is assessed for each website before and after applying the privacy protection using an A to F scale, the positive grades appearing in green and the negative ones in red. The use of graphic elements helps navigating through the options. Its features are: integrated ad-blocker and private search activity. SafeShepherd is an accounts' cleaning tool with a simple interface; there are several customization possibilities, the navigation and configuration is intuitive, but there are a lot of information to read. Its features are: alerts, privacy monitoring, and a personal privacy expert. All four apps offer similar functionalities and tend to privilege simple and user-friendly interfaces, as already noticed in the productivity category.

The two unplug applications, despite being different from the other applications in the sense that they are mainly tracking activities to promote behavioral change by the users, end up having the same functioning design strategy based on simplicity. Checky offers a simple and visual interface; there is not much to customize or to explore since it is a really simple app. It offers two features: the mobile phone checking counter and the map that allow users to see where they check the phone more often. Binky being a feed of random visual content is very appealing and its interface is simple since there are not many customization options. Despite not being a social media platform, it looks like one and its features are similar to those offered by this type of platforms: comment, re-bink, like, and discard/love content.

In order to grasp how mainstream media are discussing these types of tools, an exploratory textual analysis of news reports was performed. We searched for reports about the applications using the Google news filter for English language publications in 2019. The searched keywords were the names of the apps plus "time." The name of the apps were included because the goal was to specifically understand how news media have been communicating these particular applications. The main variables of this analysis were the type of media and the type of article (see Table 10.2).

Table 10.2. Media Discourses' Summary Table

App	*Media*	*Title*	*Date*	*Author*	*Type of Media*	*Type of Article*
Cold Turkey Blocker	Furniture News	Take a break from business with the latest tools	10.19	not signed	Print and digital specialized magazine	Opinion, suggestions for business owners on holidays
Cold Turkey Blocker	Business Insider	These internet-blocking apps keep me from getting distracted on the job — and they've become essential tools in my work arsenal	10.04.19	Christopher Curley	News website	Opinion, comparison between different technological solutions
Cold Turkey Blocker	RealLeaders	9 ways to deal with project diversions — and do your best work	12.09.19	Charlie Gilkey	Print and digital business and leadership magazine	Opinion, tips for time management and distractions avoidance
Cold Turkey Blocker	Inc.	3 simple steps to spend less time on Facebook and other time-wasting social media sites	05.03.19	Christina DesMarais	Print and digital magazine	Opinion, time management tips
Cold Turkey Blocker	The National Interest	Our smartphone addiction is killing us: here's how to stop it	31.05.19	Ashley Whillans	Print and digital magazine	Ppinion article about technology use, mobile phones addiction and how to control time management again
Deseat.me	Android Authority	How to delete yourself from the internet	23.07.19	Scott Adam Gordon	Digital media outlet focusing on Android operating system	Full-length opinion article about the data that is collected about internet users and how to erase oneself from the web, it presents different apps that may be helpful

(continued)

Table 10.2. *(Continued)*

App	*Media*	*Title*	*Date*	*Author*	*Type of Media*	*Type of Article*
Deseat.me	Wired	How to delete your phone's zombie apps and old online accounts	24.08.19	Matt Burgess	Print and digital magazine	Full-length opinion article about data that is tracked online from users' unused accounts and how to delete them; deseat.me is one of the tools proposed
Deseat.me	10 daily	Can you bust the ghosts of your social media past?	12.05.19	Antoinette Lattouf	News website	Article about erasing information from social media and cleaning apps, such as Deseat.me
Deseat.me	Graham Culley / Smashing Security	Smashing Security #131: Zap yourself from the net, and patch now against BlueKeep	06.06.19	Graham Cluley, Carole Theriault	Computer security news, advice and opinion blog	Podcast episode about erasing oneself from the web; one of the tools discussed is Deseat.me
Disconnect	Engadget	10 helpful everyday apps you need to know about	30.10.19	StackCommerce	Technology blog	Opinion, comparison between different productivity apps
Disconnect	The Washington Post	How to limit iPhone app tracking	28.05.19	Geoffrey A. Fowler	Print and digital newspaper	Opinion article providing tips to control tracking in mobile phones, particularly iPhones
Disconnect	The Seattle Times	While you're sleeping, your iPhone stays busy. Here's what is happening and how to limit app tracking	01.06.19	Geoffrey A. Fowler	Print and digital newspaper	Full-length opinion article about mobile phones tracking, providing a list of tips to prevent that from occurring
Disconnect	PCWorld	Disconnect Premium review: A worthy VPN that won't blow your mind	05.02.19	Ian Paul	Print and digital computer magazine	Review of Disconnect premium service

Disconnect	The New Daily	'Fingerprinting' to track us online is on the rise. Here's what to do	08.09.19	Brian Chen	Digital newspaper	Full-length opinion article about privacy and tracking online, it presents tips on what one can do and presents Disconnect as part of the solution
Disconnect	The Wall Street Journal	You give apps sensitive personal information. Then they tell Facebook.	22.02.19	Sam Schechner	Print and digital newspaper	Full-length article about data tracking using mobile phone apps
DuckDuckGo	CNBC	Companies shouldn't keep "honeypots of data" that attract bad actors, says executive at Google search rival	29.10.19	Lauren Feiner	Television business news channel and website	Article about the storage of digital data and how this puts users' privacy at risk; DuckDuckGo is discussed as an alternative that respects users more
DuckDuckGo	ghacks.net	DuckDuckGo adds Indian servers and new options	16.10.19	Martin Brinkmann	Technology blog	Article about DuckDuckGo
DuckDuckGo	PC Mag	One in five people have killed a social media profile for privacy	10.10.19	Eric Griffith	Digital and print magazine	Article about the results of a study commissioned by DuckDuckGo that presents tools that protect users privacy
DuckDuckGo	Forbes	DuckDuckGo research makes bullish case for Bitcoin and cryptocurrency	07.10.19	Kyle Torpey	Digital and print magazine	Article about the results of a study commissioned by DuckDuckGo
DuckDuckGo	Neowin	DuckDuckGo adds new servers in India, improves search, and refines dark theme	15.10.2019	Paul Hill	Technology industry news website	Article about DuckDuckGo

(continued)

Table 10.2. *(Continued)*

App	*Media*	*Title*	*Date*	*Author*	*Type of Media*	*Type of Article*
DuckDuckGo	The Mac observer	DuckDuckGo survey shows people taking action on privacy	03.10.19	Andrew Orr	Apple-related news website	Brief news about DuckDuckGo study results
DuckDuckGo	9to5 Mac	Comment: A news focused Apple Start Page would be a welcome upgrade for Safari	20.10.19	Bradley Chambers	Apple-related news website	Opinion article about what could be improved in Safari browser and one of the suggestions is to have DuckDuckGo built-in
DuckDuckGo	The New York Times	10 tips to avoid leaving tracks around the internet	04.10.19	David Pogue	Print and digital newspaper	Full length article about the information that is tracked online and how one can minimize it
DuckDuckGo	ComputerWorld	How and why Apple users should switch to DuckDuckGo for search	18.07.19	Jonny Evans	Print and digital magazine	Opinion article about DuckDuckGo
DuckDuckGo	The New York Times	A feisty google adversary tests how much people care about privacy	15.07.19	Nathaniel Popper	Print and digital newspaper	Full length article about DuckDuckGo
FocusMe	RealSimple	The best time management apps to help you work smarter, not harder	01.05.19	Maggie Seaver	Print and digital magazine	Opinion, comparison between different time management tools
FocusMe	Black Enterprise	Four of the best employee productivity apps	10.08.19	Kiara Williams	Print and digital magazine and media outlet	Opinion, comparison between different productivity management tools

FocusMe	B2C	37 of the Best Google Chrome extensions suggested by top marketers. Read more at https://www.business2community.com/digital-marketing/37-of-the-best-google-chrome-extensions-suggested-by-top-marketers-02168270	13.02.19	Nico Prins	News and trends website targeting the business community	Opinion, extensive comparison between Chrome extensions considered to be the best in different categories, among which is productivity
Freedom	Business Insider	These internet-blocking apps keep me from getting distracted on the job — and they've become essential tools in my work arsenal	10.04.19	Christopher Curley	News website	Opinion, comparison between different technological solutions
Freedom	Fast Company	Apple tight-lipped on removal of Freedom and other content-blocking apps	20.09.19	Steven Melendez	Print and digital business and technology magazine	Focus on the blocking by Apple, brief presentation of some content-blocking apps, among which is Freedom
Freedom	Gizmodo	The best apps for spending less time on your phone	17.09.19	David Nield	News blog	Opinion, comparison between different time management mobile phone apps
Freedom	lifehacker	Freedom's new Chrome extension forces you to pause before opening a distracting site	30.05.19	Nick Douglas	Blog about life hacks and software	Brief opinion article about the launch of Freedom's Chrome extension

(continued)

Table 10.2. *(Continued)*

App	*Media*	*Title*	*Date*	*Author*	*Type of Media*	*Type of Article*
Freedom	WindowsReport	Block online gambling craving with these anti-gambling tools	08.07.19	Ivan Jenic	Website focusing on Windows content - news, tips and advices for PC owners	Opinion, comparison between different apps
Freedom	The New Yorker	What it takes to put your phone away	22.04.19	Jia Tolentino	Print and digital magazine	Full-length article about mobile phone addiction and how to control it
Freedom	The New York Times	Ready. Set. Write a book.	30.10.19	J. D. Biersdorfer	Print and digital newspaper	The article is about writing; this type of app is just mentioned as a tool for time management
RescueTime	The New York Times	Ready. Set. Write a book.	31.10.19	J. D. Biersdorfer	Print and digital newspaper	The article is about writing; this type of app is just mentioned as a tool for time management
RescueTime	PC Mag	5 ways to cut back on social media	07.10.19	Jill Duffy	Print and digital computer magazine	Article about social media usage management, opinion, suggestion of different apps to control different aspects; RescueTime is proposed for time management

RescueTime	Furniture News	Take a break from business with the latest tools	10.19	not signed	Print and digital specialized magazine	Opinion, suggestions for business owners on holiday
RescueTime	Thrive Global	5 lifestyle apps that make you more productive	01.08.19	Michael Usiagwu	Content website focusing on how technology and media may improve our quality of life	Opinion, comparison between productivity apps
RescueTime	Refinery 29	5 productivity apps you definitely want to try	25.05.19	Refinery 29 Editors	Digital media outlet targeting young females	Opinion, comparison between 5 productivity apps
RescueTime	TechRadar.pro	Best time management solution in 2019: Apps and software for time tracking	19.09.19	Jonas DeMuro, Brian Turner	Technology news and reviews website	Opinion article about time management software, comparison
RescueTime	RollingOut	6 great apps for help with time management	20.04.19	Rolling	News website	Brief comparison between different time management apps, opinion

We found seven articles that mentioned Freedom and its features. The majority are opinion articles and are focused on comparing different solutions to control time spent online. The majority of the sources are digital media outlets. Rescue Time also counts seven articles, being the great majority comparisons between productivity, time management apps; the sources are mainly digital media outlets. Cold Turkey Blocker has five articles, all opinion articles mainly focusing on tips for a better time management. FocusMe, on the other hand, was mentioned in three articles, only one published by a digital-only media, all are opinion articles, presenting comparison between different apps. Disconnect counts six articles, diverse genres, four full-length opinion articles, one review and one brief opinion article. The longer articles were concerned with mobile phones tracking and Disconnect is presented as part of the solution to control this situation. Deseat.me has three articles and one podcast episode. All articles are about deleting personal information from the web and Deseat.me is presented as one of the cleaning services one can use. DuckDuckGo is the app that counts with the highest number of articles, ten tackling different topics; four are centered on DuckDuckGo, three on the results of a study commissioned by DuckDuckGo, two about the information that is tracked online and how one can minimize it, and one opinion article about iOS browser Safari and how it could be improved. The majority of the sources are legacy media ones that also have a digital news website. The other three apps, SafeShepherd, Checky, and Binky, returned no results for news published in 2019. The most popular applications in terms of user numbers tend to be those that get more media coverage, DuckDuckGo being the only exception. Digital media outlets tend to focus the attention in comparisons between apps presented in short articles, while legacy-digital media outlets tend to invest in full-length thematic articles. The apps that had more media attention are those that are related to questions considered to be of great importance, such as users' privacy online.

CONCLUSION

As social media platforms have evolved and are widely used by internet users, fulfilling many different needs ranging from the more general to the more specific and personal social needs, they tend to become more sophisticated, making available new features aiming at contributing to users' engagement and loyalty. How interfaces are strategically designed is posing challenges to user's rights. More transparency of the mechanisms is necessary including ethical guidelines for the design of interfaces. One counterpoint to these strategies has been the emergence of technological solutions that aim at reaching balance in the time spent online.

It is interesting to note that the analyzed apps do not focus on "disconnecting" the users but to disconnect them from intrusive content in order to assure that their productivity levels remain high or that their privacy is safeguarded. Most of the discourses of and features made available by these apps focus on having time to do more. None of the analyzed cases focused on having more time for leisure activities, or to be with the family, for instance, and on users' right to control and manage their own digital experiences. There are even apps that claim to promote a healthy relationship with the mobile phone, but that may end up reinforcing the maintenance of frequent mobile use practices. This follows Han's (2015) claim that society is pushing for disconnection as a form of being active digitally.

With such a huge impact on the everyday life of millions of users around the world, social media should be under scrutiny and follow a strict ethical practice. Alternative models can be thought where participation is stimulated, following ideas of co constructive or participatory design of social media that goes beyond self-regulation

REFERENCES

Arendt, H. (1958). *The Human Condition.* Chicago and London: University of Chicago Press.

Castells, M. (2008). Afterword. In J. Katz (Ed.), *Handbook of Mobile Communication Studies* (pp. 447–51). Cambridge, MA: MIT Press.

———. (1996). *The Rise of the Network Society* (Vol. 1). Cambridge and Oxford: Blackwell.

Chandler, D., and Reid, J. (2016). *The Neoliberal Subject: Resilience, Adaptation and Vulnerability.* Lanham, MD: Rowman & Littlefield.

Chayko, M. (2008). *Portable Communities. The Social Dynamics of Online and Mobile Connectedness.* Albany: State University of New York.

Couldry, N., and Hepp, A. (2016). *The Mediated Construction of Reality.* Cambridge: Polity Press.

Couldry, N., and Mejias, U.A. (2019). *The Costs of Connection.* Palo Alto, CA: Stanford University Press.

Cramer, F. (2015). What is "post-digital"? In Berry, D. M., and Dieter, M. (eds.), *Postdigital Aesthetics: Art, Computation and Design*, pp. 14–26. London: Palgrave Macmillan.

Deterding, S., Khaled, R., Nacke, L. E., and Dixon, D. (2011). Gamification: Toward a definition. In *CHI 2011 Gamification Workshop Proceedings*, Vancouver, BC, Canada, available at http://gamification-research.org/wp-content/uploads/2011/04/02-Deterding-Khaled-Nacke-Dixon.pdf.

Ganito, C. (2016). *Women Speak. Gendering the Mobile Phone.* Lisbon: Universidade Católica Editora.

Ganito, C., and Jorge, A. (2018, October 18–21). "On and off: Digital practices of connecting and disconnecting across the life course." Paper presented at the Selected Papers of #AoIR2017: The 18th Annual Conference of the Association of Internet Researchers. Tartu, Estonia

Giddens, A. (1990). *The Consequences of Modernity.* Cambridge: Polity Press.

Guyard, C., and Kaun, A. (2018). "Workfulness: Governing the disobedient brain." *Journal of Cultural Economy, 11*(6), 535–48.

Han, B-C. (2015). *The Scent of Time: A Philosophical Essay on the Art of Lingering.* Cambridge: Polity.

Hesselberth, P. (2018). "Discourses on disconnectivity and the right to disconnect." *New Media & Society, 20*(5), 1994–2010.

Ito, M., Okabe, D., and Anderson, K. (2008). Portable objects in three global cities: The personalization of urban places. In R. Ling and S. W. Campbell (Eds.), *The Reconstruction of Space and Time: Mobile Communciation Practices* (pp. 67–85). New Brunswick, London: Transaction Publishers.

Johnsen, T. (2001). "The instantaneous time: How being connected affect the notion of time." Paper presented at the IT-Users and Producers in an Evolving Sociocultural Context.

Karlsen, F., and Syvertsen, T. (2016). "You can't smell roses online: Intruding media and reverse domestication," *Nordicom Review*, 37(special issue), 25–39.

Katz, J., and Aakhus, M. (2002). *Perpetual Contact: Mobile Communication, Private Talk, Public Performance.* Cambridge: Cambridge University Press.

Kaun, A., Hartley, J. M., and Juzefovičs, J. (2016). "In search of the invisible (audiences)." *Participations*, 13, 334–48.

Kaun, A., and Schwarzenegger, C. (2014). "'No media, less life?' Online disconnection in mediatized worlds," *First Monday*, 19(11). http://firstmonday.org/ojs/index.php/fm/article/view/5497/4158

Light, B., and Cassidy, E. (2014). "Strategies for the suspension and prevention of connection: Rendering disconnection as socioeconomic lubricant with Facebook." *New Media & Society*, 16, 1169–84.

Ofcom. (2017). Communications Market Report. https://www.ofcom.org.uk/__data/assets/pdf_file/0024/26826/cmr_uk_2016.pdf,

Portwood-Stacer, L. (2012). "Media refusal and conspicuous non-consumption: The performative and political dimensions of Facebook abstention." *New Media & Society, 15*(7), 1041–57.

Rosa, H. (2003). "Social acceleration: Ethical and political consequences of a desynchronized high-speed society." *Constellations, 10*(1), 3–33.

Southerton, D. (2006). "Analyzing the temporal organization of daily life: Social constrains, practices and their allocation." *Sociology, 40*(3), 435–54.

Southerton, D., and Tomlinson, M. (2005). "Pressed for time"—The differential impacts of a "time squeeze." *The Sociological Review, 53*(2), 215–40.

Statista. (2009). Online seach usage—Statistics & facts. Retrieved December 17, 2019, from https://statista.com/topics/1710/search-engine-usage/.

Syvertsen, T., and Enli, G, (2019). "Digital detox: Media resistance and the promise of authenticity." *Convergence*, 1–15. https://doi.org/10.1177/1354856519847325.
Thorén, C., Edenius, M., Lundström, J. E., and Kitzmann, A. (2019). "The hipster's dilemma: What is analogue or digital in the post-digital society?" *Convergence 25*(1), 324–39.
Van Dijck, J., and Poell, T. (2013). "Understanding social media logic." *Media and Communication, 1*(1) 2–14.
Wajcman, J. (2008). "Life in the Fast Lane? Towards a Sociology of Technology and Time." *The British Journal of Sociology, 59*(1), 59–77.
Woodstock, L. (2014). "Media resistance: Opportunities for practice theory and new media research." *International Journal of Communication, 8*(19), 1983–2001.

Index

Note: Page references for figures and tables are italicized.

About the Editors and Contributors

Alex Beattie is a PhD candidate at Te Herenga Waka–Victoria University of Wellington. His thesis is titled *The Manufacture of Disconnection* and assesses technology-based ways to disconnect from the internet. His research interests include disconnection, digital well-being, persuasive design, and Silicon Valley culture.

Hannah Ditchfield is a research associate in the Department of Sociological Studies at the University of Sheffield. Her research is centered on digital media and the everyday and has included work on digital interaction, online identity, and social media affordances. She is currently working on a project funded by the Nuffield Foundation, "Living with Data," which aims to understand people's knowledge, experiences, and perceptions of data practices and what would make them "fair." To date, her research has been published in *New Media & Society* and *The SAGE Handbook of Qualitative Data Collection*.

Mikolaj Dymek is a senior lecturer at the Department of Media Technology at Södertörn University, Stockholm, Sweden. Previously of Royal Institute of Technology (Stockholm), Uppsala University and Mid Sweden University, he researches the intersection of marketing communications, digital media, and game studies. He has published in journals such as *Consumption Markets and Culture*, *Contemporary Social Science*, *Ephemera: Theory and Politics in Organization* ,and others. He has also co-authored *Video Game Marketing* and edited *The Business of Gamification*, both published by Routledge.

Cátia Ferreira is an assistant professor in media and communication studies at the Human Sciences Faculty (FCH) of Universidade Católica Portuguesa (UCP), Portugal. She is coordinator of the BA in social and cultural communication and of the postgraduate course Communication and Content Marketing. She is a senior researcher at the Research Centre for Communication and Culture (FCH, UCP), being also part of the scientific board, and a researcher at the Centre for English, Translation, and Anglo-Portuguese Studies (FCSH, UNL). Her areas of research and teaching are digital media, particularly digital games, social media and mobile devices, transmedia storytelling, multimedia communication, and digital reading practices.

Roxana Moroşanu Firth is a Research Fellow at the Centre for Computing and Social Responsibility, De Montfort University, United Kingdom. Her research looks at people and technology, with a focus on smart homes, innovation, and creativity. She is the author of *An Ethnography of Household Energy Demand in the UK* (2016) and co-author of *Making Homes: Anthropology and Design* (2017). Her research has appeared in, among others, *The Cambridge Journal of Anthropology*, *Design Studies*, *International Journal of Cultural Studies*.

Ingrid Forsler is a senior lecturer in media and communication studies at Södertörn University in Stockholm, Sweden, where she recently defended her PhD thesis about the relationship between media and visual art education in Sweden and Estonia. She has a background in art and media education, and her research interests include media literacy education, creative research methods, visual culture, media materiality, and media infrastructures.

Carla Ganito is an assistant professor in media and communication Studies at the Human Sciences Faculty (FCH) of Universidade Católica Portuguesa (UCP), Portugal. She coordinates the research group DIGLIT—Digital Literacy and Cultural Change of CECC—Research Centre for Communication and Culture. Her areas of research and teaching are digital media with a focus on mobile media, gender and technology, and cyber culture.

Carina Guyard is senior lecturer in media and communication studies at Södertörn University, Sweden. Her research has focused on gaining a deeper understanding of what it means to work in a corporate culture that is increasingly dominated by digital technologies. Currently, her research interest is directed to the emergent field of "neuroeducation," and in particular how insights from neuroscience influences notions of digital media and young peoples' learning abilities.

Martin Hand is associate professor of sociology at Queen's University, Kingston, Canada. His central research trajectory has been the unintended social and cultural consequences of digitization in everyday life. His previous publications include *Big Data?* (co-edited, 2014), *Ubiquitous Photography* (2012), *Making Digital Cultures* (2008), *The Design of Everyday Life* (co-authored, 2007), plus articles and essays about visual culture, photography, digitization, technology, and consumption.

Magdalena Kania-Lundholm is a senior lecturer in sociology at the School of Health and Social Studies, Dalarna University, Sweden. Her research combines sociology of communications and media, cultural sociology, critical internet studies, social theory, and qualitative methods. She focuses on, among others, the questions of digitalization, technology (non)use among older people, and digital inclusion. Magdalena's work was featured in journals such as *Sociology Compass*, *Journal of Aging Studies*, *Media Culture & Society*, *Digital Media & Society*, and others.

Anne Kaun is associate professor in media and communication studies at Södertörn University, Sweden. Her research is concerned with media and political activism and the role of technology for political participation in the current media ecology and from a historical perspective. She is currently studying the consequences of automation in public service institutions. Her research has appeared in among others *International Journal of Communication*, *New Media & Society*, *Media, Culture & Society*, and *Time & Society*. In 2016, she published her book *Crisis and Critique*.

Christine Lohmeier is professor of communication at the University of Salzburg, Austria. Her research focuses on questions of identity, memory, time, and media in everyday life. Recent publications addressed family memory and qualitative approaches to research. Christine's work has appeared in *Media, Culture & Society*, *Communications and the International Journal of Media and Cultural Politics* among others. Christine is the co-editor of *Diskursanalyse für die Kommunikationswissenschaft* (2019) and *Memory in a mediated world. Remembrance and Reconstruction* (2016).

Peter Lunt is a professor of media and communication at the University of Leicester, United Kingdom, in the School of Media, Communication and Sociology. His research interests include media audience research, media policy and regulation, media and social theory. He has received support for his research from ESRC, EC, The British Council, and the British Academy. He has published more than one hundred articles and chapters. His books include

Talk on Television and *Media Regulation*, both with Sonia Livingstone. He is currently working on a book entitled *Goffman and the Media* and a book with Shuhan Chen on Chinese social media.

Tim Markham is professor of journalism and media at Birkbeck, University of London, United Kingdom. His research combines sociological and philosophical perspectives to investigate cultures of media production, consumption, and circulation. Recent books include *Digital Life* (2020), *Media and the Experience of Social Change* (Rowman & Littlefield, 2017), *Media and Everyday Life* (2017), and *Conditions of Mediation* (2017), and research articles have appeared in journals including *Media, Culture & Society*, *International Journal of Cultural Studies*, *Communication & Critical/Cultural Studies*, *Journalism Practice and Journalism: Theory, Practice & Criticism.*

Manuel Menke is an assistant professor of communication and media at the Department of Communication, University of Copenhagen, Denmark. He worked at the LMU Munich and at the University of Augsburg in Germany. He holds a PhD in communication from the University of Augsburg. His research interests comprise (theories of) social and media change, media and nostalgia, mediated memories, (digital) publics, as well as journalism research. His research has appeared in, among others, *International Journal of Communication*, *Journalism Studies*, and *Convergence: The International Journal of Research into New Media Technologies*. In 2019, his book *Medienostalgie in digitalen* Öffentlichkeiten [Media nostalgia in digital publics] was published.

Christian Pentzold is professor of communication and media at Chemnitz University of Technology, Germany. Pentzold is broadly interested in the construction and appropriation of digital media and the roles information and communication technologies play in modern society. In current projects, he looks at the public understanding of big data, humans interacting with embodied technologies, the organization and governance of peer production, as well as the interplay of time, data, and media. His work has been published in places such as *New Media & Society*, *Media, Culture & Society*, the *International Journal of Communication*, and *Communication, Culture & Critique.* He is currently finishing a book on peer production.

Sean Rintel is a senior researcher of human-computer interaction in the Future of Work group at Microsoft Research Cambridge (UK). His research explores how the affordances of communication technologies interact with language, social action, and culture. He has published in *ToCHI*, *Human-*

Computer Interaction, *Pragmatics, Discourse, Context and Media*, *Journal of Computer-Mediated Communication*, and *Human Communication Research*. He has been an associate chair for CHI, a senior editor for the *Oxford Research Encyclopedia of Communication*, and a former board member and chair of Electronic Frontiers Australia.

Christian Schwarzenegger is a senior lecturer "Akademischer Rat" at the Department of Media, Knowledge and Communication at the University of Augsburg, Germany. His research interests include mediatization and digital transformation of society and everyday life, media and memory, communication history as well as qualitative methods of communication research. His research appeared among others in *New Media & Society*, *Convergence*, *Digital Journalism*, *Information, Communication and Society*.

Abigail Sellen is deputy director of Microsoft Research in Cambridge, United Kingdom, and an honorary professor at University College London, the University of Lancaster, and Nottingham University. She has wide ranging interests in the area of human-computer interaction. Her current research focuses on taking an interdisciplinary approach to the development of tools for the future of work, as well exploring the ethics of machine learning. She has published on many topics including: health care, computer input, help systems, reading, paper use in offices, video-conferencing design, search, photo use, gesture-based input, human error, and computer support for human memory. This includes the book *The Myth of the Paperless Office* (with co-author Richard Harper), which won an IEEE award.

www.ingramcontent.com/pod-product-compliance
Lightning Source LLC
Chambersburg PA
CBHW031132270326
41929CB00011B/1591
9781538149850